WHEN SMOKE RAN LIKE WATER

WHEN SMOKE RAN LIKE WATER

Tales of Environmental Deception and the Battle Against Pollution

DEVRA DAVIS

A Member of the Perseus Books Group

Published by Basic Books,
A Member of the Perseus Books Group

Designed by Trish Wilkinson

Library of Congress Cataloging-in-Publication Data
Davis, Devra Lee.
When smoke ran like water : tales of environmental deception and the battle against pollution / Devra Davis.
p. cm.
Includes bibliographical references and index.
ISBN 0-465-01521-1 (alk. paper)
1. Environmental health. 2. Environmental induced diseases—Epidemiology.
[DNLM: 1. Smoke—adverse effects—Popular Works. 2. Environmental Pollutants—adverse effects—Popular Works. 3. Risk Factors—Popular Works.
WA 754 D261w 2002] I. Title.
RA565 .D385 2002

2002010562

02 03 04 05 / 10 9 8 7 6 5 4 3 2 1

לחיים
צבי דב בן אברהם לוי ופייגא
מרים חנה בת שמעון ופנינה
אשר דוד בן נתן ומלכה
ובניהם פייול אברם ולאה זהבה

CONTENTS

FOREWORD

I FIRST MET DEVRA DAVIS nearly a decade ago when I heard her speak at the Strang–Cornell Cancer Prevention Center, where I was director of medical oncology. Her talk opened my eyes. She provided detailed documentation of the damaging impact on living systems from dioxins, dusts, pesticides, solvents, and other toxic materials. As Chief Medical Resident at New York Hospital, Cornell Medical Center, and in my later clinical work, I have continued to work on the front lines with patients struggling with cancers, most of which cannot be explained. I will never forget the question with which she concluded her talk: "What if the same environmental exposures that have been shown to affect fish, turtles, polar bears, snails, and deer—have also damaged the human gene pool?"

This question continues to haunt me. I see more and more younger and younger patients with newly diagnosed cancer. It has also largely focused my integrative oncology practice at Weill-Cornell Medical College, as I work to come up with ways to prevent recurrences and to ensure that fewer people ever become cancer patients.

When Smoke Ran Like Water will surely be attacked by those industrial polluters who claim that we lack sufficient proof that the environment is harming humans. Nothing could be further from the truth. Over the past three decades, Devra's numerous scientific papers have outlined many of the current health issues associated with environmental pollution in a highly thoughtful, carefully considered manner. Underlying this book is a voluminous scientific literature giving abundant evidence that many pesticides and industrial pollutants and their by-products cause significant numbers of cancers and other diseases in the United States and throughout the world.

The battles against environmental degradation have often been waged by lonely figures, far from the public eye, against those with a multi-million-dollar stake in having us believe that there was no crisis—that smog and soot do not induce heart and lung disease, that lead in the air does not diminish children's I.Q.s. All the tales of environmental deception relayed in this book remain sadly relevant today, as recent disclosures of corporate deceit in many spheres makes clear.

Today the battleground over pollution has shifted to activities that put us all at risk, including carcinogens and other toxic agents, which can leave permanent but secret scars. The human body, like the planet itself, accumulates many known and suspected toxins.

We live in an era when one in three Americans will hear, at some point in their lives, the words "you have cancer." I have had to say these words myself many times, and it never gets easier. I see the suffering cancer causes every day. We wasted fifty years debating the role of cigarettes in causing cancer, and we cannot afford to waste another fifty years before we develop strategies to prevent environmental cancer and other avoidable diseases.

At the U.S. National Academy of Sciences, Carnegie Mellon University's Heinz School, the World Health Organization, World Resources Institute, and at a number of front-rank institutions in this country and abroad, Devra Davis has spent the last two decades creating a veritable blueprint for investigating and controlling environmental hazards. The first step, as this book clearly demonstrates, is to start asking some obvious questions. Here is one to start with: What we are doing to the planet, are we also doing to our own bodies?

Mitchell L. Gaynor, MD
Weill-Cornell Medical College
New York City

PREFACE

Not everything that can be counted, counts.
Not everything that counts, can be counted.
—ALBERT EINSTEIN

IN THE EARLY 1980S, I reached a disturbing conclusion. I was working at the National Academy of Sciences on what turned into a four-year-long study of air inside airplanes. The investigation didn't need to take four years, or even one. But Senator Daniel K. Inouye had given the Federal Aviation Administration half a million dollars to fund a committee at the academy to find out why he kept getting sick after his regular eight-hour trips from Honolulu to Washington, and no sane institution walks away from ready money.

I figured out an easy way to answer the senator's question. From a friend at the Environmental Protection Agency, I borrowed a clunky piece of equipment called a piezobalance, which could measure the weight of airborne particles smaller than a human hair, such as those produced by cigarette smoke. I set off on a flight to Paris carrying a ten-pound metal box that looked for all the world like a bomb, concealed under my grandmother's old mink coat. Just in case anybody got suspicious, I had an official letter indicating that I was transporting a scientific instrument. I never needed it. This was 1983.

By the end of the flight, I had the answer. The levels of particles in the smoking and nonsmoking sections were identical. The senator kept getting sick because, for all his lungs cared, he might as well have been sitting with the heavy smokers.

When I got back to Washington I eagerly told my boss at the academy the good news. "We don't need to do a study for the senator!"

He looked at me nervously and asked, "What are you talking about?"

I suggested we could save time and money if we went out and studied a couple more planes and prepared a short report.

After I explained what I'd done, he sighed and shook his head. "You can't do anything with those numbers. No committee reviewed what you were going to do. Nobody approved this project. Forget about it. We have to proceed according to our rules here."

Half a million dollars and four years later, the official academy study confirmed what I had found in a single flight. After a few hours, there really was no nonsmoking section on an airplane. Smoke made people uncomfortable, no matter where they were sitting, and it also gummed up the plane's instruments. Within the year of the publication of this report, bans on smoking in airplanes and other public places began to take hold.

Although we were able to mount a national and, eventually, an international campaign to rid airplanes of the minute residues of cigarette smoke, no campaign has ever addressed indoor exposures through agents other than cigarettes. Many toxic compounds found in cigarette smoke are still regularly inhaled throughout the country by people who, like airline passengers of old, have no choice in the matter.

Among the places where these exposures can occur are the so-called clean rooms so crucial to the global electronics revolution. To protect the wafer-thin brains of computers from microscopic contamination by human hair, skin cells, dandruff, and other minute dust, workers are covered head to toe in gauzy, white paper outfits called bunny suits. But the shielding that these suits provide goes largely one way—it protects the chips but not the workers, who may absorb into their skin, lungs, and bloodstream over forty different chemicals commonly used in high-tech manufacture. Many of these, such as benzene, asbestos, and chlorinated organic solvents, were well known by the early 1980s to cause cancer in both animals and humans. The disturbing conclusion I reached was this: Combined exposures to these toxic agents would cause more and more cancers among those with the greatest exposure over the longest time. I felt as if a hideous natural experiment was about to unfold before my eyes.

Everything people do leaves trails of numbers. Births, deaths, the sex of our children, smoking, even the size and shape of our private parts can all be measured and counted. Some of these measurements will mean nothing, but others tell who we are, how long and how well we are living, where we have been, and whether we will ever get to where we think we're going. My job is to find the right things to count and measure—to make meaningful patterns out of what everyone else thinks is chaos. The trails of numbers seldom stop: One finding may lead, quite unexpectedly, to something entirely different.

In a series of articles that appeared in the 1980s and 1990s in the *Journal of the American Medical Association, Lancet,* and other scientific journals, I had warned about the dangers of toxic chemicals to which workers are exposed. Because I had done this, one sunny morning in January 2000, I found myself seated at a mahogany table in a paneled conference room in the company of a lawyer named Amanda Hawes.

My job, as I said, is to find patterns arising from the way people live and work. Sometimes those patterns are bizarrely menacing. Besides Mandy and myself, there were eight middle-aged men arrayed around the table, each wearing a dark-gray suit of the kind of quiet ostentation that denotes extreme wealth in modern-day America. We were well paid to be there: The cost of assembling the ten of us in one place for the day was more than what the average factory worker in a clean room made in a year.

"Good morning, Doctor Davis," said the shortest and best-dressed of the group. He spoke with the ease and cadence of someone to whom time was no object, who knew that his listeners were forced to weigh his every word. "*My* name . . . is Dean Allison." He smiled and looked me straight in the eye. I forced myself not to blink.

The man next to him spoke up. "Good morning, Dr. Davis." He smiled tightly and continued, "My *name* is Dean Allison."

Then the third man said, "Good morning. My name is *Dean* Allison." At this they all broke into laughter.

These Allisons represented giant companies worth billions of dollars—Hoechst Celanese, IBM, Dupont, Dow Chemical—and they were there to defend a potentially very costly lawsuit. The plaintiffs, whom Amanda represented, were cancer sufferers—some under age forty—who claimed that their disease resulted from working under hazardous conditions that these companies had done too little to warn them of or protect them against. I was there that day to be deposed as an expert witness in the case.

The laughing introductions were no joke: The lawyers were reminding me that I was the only person in the room who was under oath. While I had sworn to tell the truth, the whole truth, and nothing but the truth, they had made no such vow and could lie, deceive, obfuscate, or pretend incomprehension whenever it suited their clients' interests. If what I said was not precisely what they wanted to hear, they could simply declare my answers "unresponsive."

I thought, naively, that my deposition would take perhaps two days. After all, we were talking about well-known carcinogens, identified as such by the World Health Organization. As a public health researcher, I am very familiar with the accepted standards of proof in my field, which rely on statistics and public health studies. Human exposures to benzene, asbestos, cadmium, and certain solvents have repeatedly been linked to increased cancer risks. I knew of hundreds of studies on animals and humans that supported this view—so many, in fact, that I hadn't even bothered to count them all. And the facts in this case fit right in. In a single work center owned by IBM, there had been six cases of testicular cancer among five thousand workers in ten years. To find six cases of that type of cancer at random in the general population, you would need two million men. The rate was even higher in a group of clean-room workers in France. For the field of epidemiology, this kind of clustering of such a rare disease constitutes a slam dunk.

The Allisons were not being paid to see it that way. They were not impressed by statistical proofs or by studies in experimental animals. Instead, they saw cancer deaths in the same terms as deaths from a handgun. When and where was the bullet fired that killed or injured each person? Who was holding the gun? They demanded that for each person who had fallen ill, we produce absolute proof that the specific worker had been exposed to specific chemicals that had caused his or her specific cancer. They wanted accounts not only of research but of bodies felled. And they were prepared to do whatever it took to keep the case going on and on. They counted on eventually wearing out the resources and abilities of the plaintiffs and their experts. My two days quickly became weeks, and by 2002, my deposition had continued, on and off, for more than two years.

Our differences, which could charitably be described as philosophical, are over how one knows what causes a disease and what one's responsibility is in light of that knowledge. Because cancer is produced by so many things that do not leave any special marks on their victims, the medical community can rarely pinpoint the causes of any one person's illness. People lead complicated lives, pay mortgages, move, marry, divorce, and do many other things that make it challenging to figure out why they become sick or die.

It would be easier if those who sustained certain exposures all turned bright colors, or if they all acquired the same patterned suits. Sometimes we put tissues and blood under the microscope to find biochemical signs and signals of what lies behind people's ailments. Sometimes neither a microscope nor a doctor is needed to figure out that something truly strange is going on.

In September 1944, a policeman in the Bowery found a ragged eighty-two-year-old man retching and collapsed on the sidewalk on Dey Street, near the Hudson Terminal in New York City. As soon as he looked at the fellow, the cop realized that this was not just a sick old drunk. The old man was sky blue. In one day, a total of eleven thin, unmarried, underemployed men had shown up at the same clinic in the Bowery in lower Manhattan, complaining of coughs, chest pains, and fatigue. All of them were blue.

If all these men had been tired and full of flu, but with normal skin tone, no one would have ever thought to connect their problems. Only their exceptional hue made the local doctors ask what was going on.

Normal skin has a rosy color because it contains iron in the form of hemoglobin. Blood saturated with oxygen is bright red. Blood that has lost its oxygen is dark bluish red. Carbon monoxide and other compounds, like those containing cyanide or some forms of nitrogen, can block iron from entering hemoglobin and deplete the blood of oxygen. People who lack oxygen in their blood tend to look blue.

The health department officials learned that every one of these blue men had become sick within a half-hour of eating at a local diner. Ten of them had oatmeal, rolls, and coffee; one had eaten just oatmeal. Attention turned to what on earth was in that oatmeal. One of the saltshakers used in preparing the cereal, it turned out, had been mistakenly filled with white crystals that looked just like sodium chloride but were actually sodium nitrite, a chemical used in the curing and preservation of meats and other foods. It tastes salty because it contains sodium, just as salt does, but it also contains nitrites that can wreak havoc in the blood. When bound to nitrite, hemoglobin cannot absorb iron and takes on an abnormal blue color, a condition called methemoglobinemia. Once this puzzle was solved, both the saltshaker and the men returned to normal.

The story of the eleven blue men provides a spectacular example of what the work of epidemiology (the word comes from the Greek *epi,* "upon," and *demos,* "people") is largely about. Epidemiologists look for common connections of patterns of illness among groups of people. Their work is easiest when exposures are knowable or controlled and take place over relatively short periods. Thus the challenge of the blue men was straightforward. They got sick quickly. They did not die. As a result, they could be questioned about what had happened to them in the days before they became ill. In fact, about 125 people ate the same oatmeal, but only these 11 turned blue and sick. The blue guys all happened to be heavy drinkers and had added extra salt to their cereal, in an effort to make up for the sodium deficit common to alcoholics.

Epidemiology works well when conducting carefully controlled, randomized trials of pharmaceuticals. Under this system, people with similar personal characteristics—age, weight, height, medical histories, and so on—are assigned to be given a carefully determined dose of a drug (or not), and information about them is rigorously charted and tracked. The two populations—subjects given the drug and those not—are compared for therapeutic effect, side effects, and other complications. All outside variables are carefully screened away, so that in the end, one can say with numerically defined confidence that *X* drug cured *Y* disease *Z* percent of the time.

But the world outside of drug trials, the world of what is called environmental epidemiology, is not such an easy or elegant place to study. People seldom come in matched groups. They almost never turn blue. And they often couldn't tell you what they've been exposed to yesterday, let alone at some critical time in their past.

The work of environmental epidemiology remains a blunt instrument. We cannot say that any one person's disease was caused by this particular exposure to this particular chemical on this particular day. The best we can say with any certainty is that if a particular chemical or group of chemicals were not in the environment, some number out of every hundred people who got sick would have remained healthy, and some number of those who died might still be alive. Is this enough? Will this style of reasoning persuade a federal court to make powerful corporations pay damages to sick plaintiffs or to their survivors? Can such an attenuated view of causation provoke anyone's righteous anger?

Consider cardiovascular disease, the most common cause of death in many industrial countries, including the United States, where half a million die from the disease each year. We know a great deal about the roles of genes, smoking, diet, and exercise in causing heart disease. But we also now know that many these deaths are at least partly environmental. In California, Brazil, England, and every other country where scientists have bothered to look carefully, they have found something quite amazing: When folks who live in more highly polluted areas have chest pains or heart attacks, their chances of dying of heart disease are nearly 30 percent greater than if they lived in cleaner regions. This is true in Riverside, California; northeast London; and São Paulo alike—the hearts of people living in polluted areas are less resilient and more vulnerable.

If you think attributing 30 percent greater risk of dying from cardiovascular disease to pollution must be an exaggeration, feel free to reduce it—say, by two-thirds, to 10 percent. You are still left with forty thousand people every year who are lost to their families prematurely from heart disease brought on by pollution. To each of these people and to their families, this is a tragedy. I believe it is a preventable one. But the pattern must be noticed and acknowledged before prevention becomes even remotely possible.

Take another example. Sudden infant death, though on the decline in the United States, still happens far more often than anyone can explain, ten thousand times each year. Most cases of crib death are painful enigmas. But the evidence is pretty clear that babies born in more polluted regions, both in the United States and elsewhere, have nearly twice the risk of dying suddenly than those born in cleaner regions. This is a classic example of how the traditional medical focus on individual diagnosis and treatment utterly fails to reveal the pattern individual cases of disease make when matched with environmental conditions. The National Center for Health Statistics can tell us how many mothers have anemia, diabetes, or hypertension. It cannot tell us how many of them live within one mile of a hazardous waste dump, drink water whose contaminants exceed recommended limits, or regularly breathe dirty air. These are not considered health statistics.

In a sense, they are not. "Environmental contamination" is never listed as the cause of death on anyone's death certificate.

The relentless American focus on health sustains popular magazines and fills the airwaves with health claims for cereals, vitamins, pills, and exercise equipment. We are told how to eat the right foods and avoid the wrong ones, practice safe sex, work out properly, get regular checkups, think healthy thoughts, and accept responsibility for our own destinies. But no matter how diligently we take care of ourselves, no matter how often we say no to unsafe sex or dangerous drugs or choose the salad instead of the hamburger, we cannot control the influences of the world around us. Where you live and work, what you eat and drink and breathe, what happened to you just before birth—all these things play critical roles in determining your prospects for health. But when illness is not a matter of personal prevention, scientists and media alike become strangely reticent.

When it comes to hazards in the workplace and environment, the safe response, which has come to be accepted as scientifically responsible, is to say nothing and do nothing until we have clear proof that the hazard has actually made people sick. When we can't marshal definitive statistical proof of a toxin's specific harmful effect, backed by a clear theory of the mechanism of that effect, it has become standard to say that we simply don't know whether the toxin is harmful or not. The absence of evidence of harm—even when no effort has been made to gather such evidence—becomes grounds for inaction. The lack of elevated levels of certain toxic chemicals in women when they are first diagnosed with breast cancer, for instance, is taken as proof that exposure to these chemicals many years earlier does not cause the disease. Of course, these earlier exposures cannot easily be reconstructed and could well be critical, but they are no longer evident. Not acting to reduce or control our use of such suspected toxic materials is a form of acting.

To whom, exactly, are we being responsible when we remain silent in the face of environmental hazards? I would like to advance another notion of responsibility, the same one we employ every day when, as parents, we send our children out with umbrellas if it looks rainy and lunch money if they will need to buy lunch. We don't wait for them to go hungry or get drenched before we acknowledge that these misfortunes should be prevented. Yet when it comes to environmental health, we are expected to wait until after the fact—until there are dead bodies or ill people to count—before taking action to prevent those and other harms from happening. Sometimes not even then.

In *When Smoke Ran Like Water,* I argue for a fundamentally new way of thinking about health and the environment. We must move beyond inflicting guilt on individuals for the things they failed to do, and instead appreciate the importance of the things that happen to all of us. Where the health of large numbers of people is at stake and the harm is potentially irreversible, it is far better to err on the side of caution. We accept this principle in many areas of public life: We do not wait for buildings to fall down or bridges to collapse before reinforcing and inspecting them for safety; we do not wait for boats to sink before requiring that they carry life jackets.

Our knowledge of the health consequences of both local and global pollution is more detailed and accurate than it has ever been. We are now in a position to make informed choices as a society about what risks we will accept and how much we're willing to pay to change them. Some have argued that a dirty world is the unavoidable price of economic growth. Those with a vested interest in not changing the causes of pollution will too often use this claptrap as an excuse for doing nothing and learning nothing.

In this book, I show some of what we've learned about the health consequences of air pollution, about contaminants and cancer, and about the planetary reach of pollution. I explain why it has taken so long to reach this point of addressing environmental concerns about human health, and I recount the struggles of those who have tried to raise these concerns. Not everybody plays fair.

There is power and beauty in learning how to reckon with numbers. Properly understood, numbers can tell us stories of the conditions that shape our lives, our deaths, and even the sex and health of our children. If we know how to listen, numbers can tell us the truth. This book tells the story of how I learned to count the patterns of life and death created by powerful forces in our modern environment.

When I was five, numbers were not important to me. I grew up in a place so small that schoolteachers did not take attendance. There were so few students in the combined first-, second-, and third-grade class that you could tell right away if someone was missing. Children could not get lost, because everybody knew who belonged to whom, and life was so plain that it seemed as if nothing could be concealed.

I loved listening to stories from the local rabbi, who taught me and two dozen other children in the one-room heder of the town's small, yellow-brick synagogue. In the summer and spring, we would sit transfixed for hours on his wooden porch as he told us enchanting tales from the Jewish tradition of Midrash, or storytelling.

One Midrashic legend has stayed with me for half a century. A group of workers is asked to do something quite difficult and complicated.

They protest, "The day is short! The work is too difficult! The project is too big! We do not have the right tools! And anyway, we are too tired! We will never finish this job!"

Their teacher replies, "It is not for you to finish the task. But—you must begin."

Part One

ANCIENT HISTORY

My Bubbe Fanne loved what we western Pennsylvanians call "rolly coasters." Every year the entire family joined our small town's annual expedition to Kennywood Park, near Pittsburgh, to ride some of the best coasters in the country. I loved the thrill of sitting securely next to my large, soft grandmother while being dragged slowly upward, and then whooshing down so fast you simply had *to scream to breathe. Once, when I was five, as we staggered off the famous Jack Rabbit with its double-dip drop, Bubbe Fanne brought me over to greet her cousin Sadie.*

She said to Sadie proudly in Yiddish, "This is my granddaughter! She is a little devil, with a big mouth!"

Though Bubbe had bragged about my ability to speak, I could not. Sadie had arrived a few months earlier from Europe, where everyone's relatives seemed to come from, and had been the subject of conversations I barely understood. Still flushed and breathless from the ride, I found myself facing a strangely silent, ghostlike woman. I don't remember if I even looked at her face: My eyes went to her forearm. On her freckled skin were some faded blue numbers, a bit deeper than the U.S.D.A. stamps on the meat my mother used to buy from the butcher. Following my gaze, my grandmother said, "Tell her, Sadie! Tell her what this means!"

I will never know what Sadie felt then, less than five years out of the concentration camps, standing there amid the calliope music and the screams of children. Like all those with tattooed numbers on their bodies, she was immune from ever having to talk about what she had lived through. She said nothing.

Other survivors swore that nobody would ever forget. One who made sure to pass on his memories of what lay behind those numbered arms was Sol Filler, the father of my friend the performance artist Deb Filler.

On August 4, 1942, Sol and his brother Tootzie had their last kiddush supper in the small Polish village of Brzozow. In their town at the foothills of the Carpathian Mountains, in the Province of Galicia, Jews were being shot in the streets. The family prepared to say good-bye. The two brothers meant to run away into Russia that night with some teenage friends. Their plans were thwarted when the mayor, a Nazi collaborator, knocked on their door during the meal and demanded that their father produce bread for the army. "Either you bake four hundred loaves of bread tonight," the mayor told them, "or I will shoot you myself."

The brothers stayed; the bread was baked. Their friends who ran away were never heard from again. The Nazis murdered all the residents of Sol's village, except three: fourteen-year-old Sol, his brother, and his father. They were spared because they ran the town bakery. The Nazis needed bread more than they needed three more dead Jews.

At Auschwitz, Sol went through another numbing process of selection. Out of twenty-five thousand persons who arrived when he did, all but five hundred were gassed. Having the skill of baking, Sol was picked for work. He survived until the winter of 1945. That January, as the Germans retreated before the Soviet Army, they forced thousands to march five hundred kilometers westward through bone-chilling cold to the remnants of their phony model work camp at Tereisenstadt. Starving and in rags, Sol understood that as long as he could keep moving, he had a chance of staying alive.

When the war ended, he landed at a displaced person's camp in Dachau. He said he began to feel like a human being again only in the winter of 1946, when he heard Leonard Bernstein perform George Gershwin's Rhapsody in Blue.

Some forty years later, Sol and Deb trekked through his remarkable path of survival. During their journey they encountered one of the frequent tour groups at Tereisenstadt. "About twenty thousand people left Auschwitz to march here," said the guide with well-rehearsed authority. "A lot of them survived. The Germans kept meticulous records on everyone who ever came here. . . . Here was the torture chamber."

Much to the woman's surprise, Sol objected. "'Scuse me. That is not true! How could you say such a thing? Nobody was writing down the numbers. I came here from Auschwitz with only two hundred. We started with ninety thousand. We were eating grass along the way. Nobody was counting the bodies. For three days we sat in this cell with no food. Bodies were all around us. If I had died here, nobody would have known."

The Holocaust was one of the great unmentioned shadows over my childhood. The other was pollution. The two are not morally comparable. If by mentioning them together I seem to equate them, reader, please be assured that is not my intention and I understand the differences. Unlike the Nazis' infamous killing machines, with their explicit plans and strategies of humiliation and murder, the sources of death and injury from pollution cannot easily be pinned down. The damages tied with our environment arise not from some deliberate effort to purify society against a perceived enemy within but from the daily activities of people doing productive work. In my family and in my community, the environment and the Holocaust had one thing in common: Both subjects were so grim and hard to stomach that they were not fit for normal conversation. Still, there is a temptation—at least among some—to accept the consequences of pollution as necessary, to submit more readily to the evil of denying or lying about it, and to excuse those who would have us not count its effects. As a result, those felled by environmental conditions seldom even know why they are dying.

The Talmud says, whoever saves a single life, it is as if she has saved the entire world. Like Sol Filler, I have come to believe it is a moral necessity that each fate be counted. That the Nazis did not track all those they murdered compounds the obscenity of their crimes. We cannot change the past, but we can and must bear witness so that the future may be different. It makes no difference whether we are remembering millions, or a few felled by a hidden, local epidemic. Those of us who survive must enumerate, count, tally, and measure what has happened. We must record and pass on the truth.

1

WHERE I COME FROM

The Lord giveth and the Lord taketh away,
but He is no longer the only one to do so.
—ALDO LEOPOLD

DONORA, PENNSYLVANIA, was the kind of place where an adventurous three-year-old like my brother Marty could wander five miles away from home and never really be lost. He made front-page headlines both times he did it—"Runaway Marty Does It Again!" read the second one. Each time, somebody brought him back up the steep hills, around the curvy, slag-lined, coal-paved roads, back to our house.

All of us children roamed free. Behind my house was a barren stretch of caked, light brown earth the size of two football fields. At its edge, the smooth, dusty ground sloped down, at an angle perfect for sliding, to some black ditches with iridescent pools of oily water at the bottom. After a few hours of playing with my friends in the fantastic cracks and crevices of this field, I usually found myself half a mile down the road, at my beloved grandmother Bubbe Pearl's house. She was always home, and her bedroom was perfumed with the smell of chicken soup.

Nestled into the hill inside a sharp horseshoe bend in the Monongahela River, Donora had sprung up around its metalworks and steel industry. In 1900, William Donner began building an iron mill alongside the fast-moving river, and enough immigrants showed up for jobs that the town was officially incorporated a year later. By midcentury it featured a church or two on most corners, an intense Little League system and one of the best high-school football teams in the valley. The main street ran for two blocks with no traffic light and was anchored at one end by the Fraternal Order of Eagles, the Masons, the Polish Falcons, the Sons of Croatia, and a bowling alley. An ice-cream cone with two big scoops cost a nickel at Weiss's Drugstore, and at Niccolanco's, a single penny could buy a child's fortune in sugar: five Tootsie Rolls, three red-hot jawbreakers, or ten smaller gum balls.

Nobody needed a clock. Dinner times, school recesses, and PTA meetings were announced by the shrieking mill whistles. When there was a fire, long blasts from the mill would signal what precinct of town the fire was in. Short blasts would indicate the street number. Any time a fire whistle went off, anyone who could stopped whatever he was doing to go help. The firemen were all volunteers.

Everybody in Donora either worked for "the world's largest nail mill," as the sign atop the factory gate announced, or worked to feed, clothe, fuel, or take care of those who did. After a full day in the factory, men would go to work on their second jobs: building the family home. First they would dig a basement, pour a concrete floor, and line it with cement blocks; then they would live there until they had saved enough to buy materials for another floor. Those who made it to the elite rank of machinist, master molder, or welder could build three stories, enough for their kids, their parents, their grandparents and usually some newly arrived cousins from central Europe. Sometimes people lived in their basements for years after they had finished other floors, simply because they could not bear to get the upstairs dirty.

Our house had only one story above the basement—cement blocks covered with pastel green aluminum siding. In the half of the basement that wasn't garage sat two huge steel sinks. When they were full of bubbles of Tide (or occasionally stronger stuff) to wash off the dirt and grease from my various expeditions, I would happily plunge into them. In those days before dryers, wet clothes were wrung through mechanical rollers to squeeze the water out into the tubs. In winter, the basement became a huge network of clotheslines, pins, and clothing in various stages of drying-out. Water from the fields turned the garage into an indoor wetland pool in springtime and sent the clothes outside to dry. Upstairs, in the tiny living room that was hardly ever used, we had thick plastic slipcovers on the furniture. The slipcovers never came off, even for company. They had to protect the precious light fabrics underneath, and they had to be wiped down every day.

Donora was a simple town, not pretty in any conventional sense, with cobblestone streets that snaked up and down hills so steep they had stairs instead of sidewalks. It was a young place, full of working people, few of them over sixty—the sort of place where weeks would pass and nobody would die.

In the 1950s, the mills began to shut down, and Donora became a place to leave. Nobody spoke about what was happening. My family moved to Pittsburgh when I was ready to begin high school, searching, like half the town's families, for better opportunities. One day I came home from classes at the University of Pittsburgh, dropped my books in the hall, and said to my mother, "Mom, was there *another* place called Donora?" I had never heard much discussion about where we came from. Now it had grabbed my attention in an unsettling way.

My mother had just put a kettle on to boil. "Why do you ask?" she said.

"Well, there are several Allentowns, several Websters, a couple of Eagles. There's Pittsburg, Kansas, and Pittsburgh, Pennsylvania. So maybe there are a few Donoras?"

She moved into the kitchen and sat down on the bench next to the built-in white table. I followed her in, took a big breath and continued to press. "I read in a book at school that in a town with the same name as ours, there was pollution. Was Donora polluted? Or was there *another* Donora?" I could not imagine that what I'd read had anything to do with where I'd grown up. I had never heard about our town being anything other than a wonderful place. I had never heard of pollution. The word sounded dirty, something to be ashamed of.

The whistle of the teakettle interrupted, and my mother got up to take the pot off the stove. At first I thought she was going to tell me about someplace else, another Donora somewhere. I was pretty sure of that, but then I could tell she was hesitating. Slowly she poured the steaming water into a small blue cup, dunking the tea bag in briefly. Without even asking if I wanted any, with a nod that commanded me to join her, she poured another blue cup of water, passed the same tea bag into it and handed it to me. We sat across the table from each other with steaming teacups between us. She sighed and finally replied, "Nobody knew from pollution then. That was just the way it was. We didn't think much about it.

"Remember all that grime we had on the cars, how we had to drive with the headlights on in the afternoon? How the sun didn't shine for days at a time? Remember how women always had their curtains hanging out to dry every week? A lot of us gave up on curtains altogether. Venetian blinds were better, because they could be wiped down. My mother's house had thirty-six windows, and we were always washing them. By the time we got to the last one, the first was already soiled. They were never really clean." I had expected an explanation, but what she gave me was a reminiscence.

There in the sunny kitchen of our big house, ten years and thirty miles away from our old town, it felt like we were on another planet. Outside I could see the sunlight on the green grass.

Bubbe Pearl never made it out of Donora. She had once been famous for her strength—the first woman in the valley to hand-crank a Model T Ford. A legendary driver, she frequently drove the nine hours to Atlantic City with her five children in tow, long before there was a Pennsylvania Turnpike. Nobody ever passed her. But when I was growing up, she kept her bed in her dining room because she could not make it up the stairs to a bedroom. She could never be more than a few steps away from an oxygen tank. Traveling beauticians regularly attended to her and to dozens of other women who were too sick to walk up and down the hills to the beauty parlor. When I was very young, I simply assumed that all blue-haired grannies stayed in bed, tethered to oxygen tanks.

"But they say people got really sick in Donora. Did people get sick?"

"Well, we used to say, 'That's not coal dust, that's gold dust.' As long as the mills were working, the town was in business. That's what kept your Zadde and your father employed. Nobody was going to ask if it made a few people ill. People had to eat."

I shot her the kind of skeptical look that daughters have been giving mothers since time immemorial.

"Look, today they might call it pollution," she sighed. "Back then, it was just a living."

So Donora was famous, but no one ever talked about it. We lapsed into silence.

Many years later I visited the remote city of Xinji, famed for eight centuries as the center of leather making in northeastern China. With some colleagues from Beijing Medical University, I was serving as an expert adviser for the United Nations Development Program. The mayor and his city planners proudly showed our delegation around a dazzling display of construction for an entire new town. The years of pollution had left the old industrial center so undeniably putrid that the whole town had recently picked itself up and moved ten miles upwind. As we toured vast areas of new construction, foundations and steel girders, nobody talked about why all this was happening.

One morning I awoke at five o'clock to the sounds of explosions. Still in my nightclothes, I rushed to the open balcony of the still-unfinished hotel where we were staying, clutching my video camera to capture whatever was going on. My colleagues rushed out as well and looked on in horror at the scene below. Smoke billowed everywhere. Workers scurried among simmering vats of black tar as a Feng Shui master set off repeated charges to rout evil.

My Chinese hosts were embarrassed by the noise and stench, but I found myself strangely exhilarated, almost nostalgic. Xinji smelled like home.

Every child in Donora knew how to make steel. You needed limestone, coal, and iron ore. A pamphlet handed out at one of Donora's American Steel & Wire Works' annual open houses explained that a normal day's operation required forty-five carloads of iron ore, forty cars of coke, six of limestone, and six of miscellaneous materials. Each day, the plant burned as much coal as did all the homes in Pittsburgh.

These ingredients regularly arrived via massive coal-fired barges snaking up the Monongahela River. Along the Donora side of the river, we could watch the barges rising through the intricate system of locks. Huge metal gates would open, the giant vessel would slowly move inside as if being swallowed by some gigantic whale, and then the gates would bellow with the crunching, creaking, groaning sounds of metal on metal as they majestically swept shut. The captain would tie up to the side of the lock with oily, blackened hawsers as thick as my leg, crossing them at bow and stern. The lock master and barge captain would wave a thumbs-up, and hundreds of thousands of gallons of muddy river water would surge into the lock. Then, with a movement that never ceased to amaze us, the ship would gradually inch upward, as though lifted by some phantom force, until it could float out the other side and continue its journey to the mills.

Other supplies came on long freight trains that ran along the river and right through the center of the string of furnaces, rolling mills, and smelters. Still others came right out of the ground nearby. Cliffs of limestone were regularly sliced away with huge shovels, draglines, and half-tracks. Family mines, some in people's backyards, yielded Appalachian coal from some of the richest seams in the world.

More than anything else, coal was essential to keeping Donora alive. It heated our homes and fired the massive furnaces and ovens of the mills. Mountainous piles of coal at the mills meant the town was in business. In addition to needing coal for the furnaces, steelmaking depended on a derivative of coal called coke. Coke is essentially coal with the greasy impurities baked out at hellish temperatures.

As a blacksmith hammers a piece of wrought iron to shape it, he must keep it hot so that it remains soft. In charcoal-fired forges like those in Donora, carbon solids and carbon monoxide remain in contact with the iron surface at relatively high temperatures. The hammered surface combines with small amounts of carbon (the iron is carburized) to create a new alloy. When it contains around one part per thousand of carbon, iron is not ordinary iron any more; it becomes steel. This small trace of carbon distributed throughout the dense mass of iron makes it stronger, so that it will take a better edge, build a stronger bridge, or support a taller building than almost any other material humans know how to make.

A coke oven in 1950 was a pretty simple affair, a gigantic beehive about the size of a one-car garage, built in honeycomb fashion out of fired bricks. Coal was shoveled in and heated to intense temperatures; coke came out. The gases and smoke that were baked out of the coal were supposed to remain completely in the oven, but they did not. Seductively sweet aromatic hydrocarbons fill the air and ground nearby.

A commercial coking operation required a string of about eighteen ovens, called a battery. Like a great shark that has to keep moving to stay alive, a coke battery had to run all the time, at temperatures above 2,000 degrees Fahrenheit. The ovens had to be blocked shut to assure a constant, even temperature. If they ever cooled, they could not be restarted because the bricks crack below 800 degrees. This meant that once the oven was fired, hardy souls with a good tolerance for heat had to carefully stack bricks together over the opening to keep the temperatures up. Folks who worked the ovens tended to be young.

By the time the workers had finished loading up and firing the last oven, the first one, having been fired the day before, was ready to yield just the right stuff for making the best steel. The oven doors would be opened on both sides, letting air into the chamber. In an instant, the air-starved coke sucked up oxygen and exploded with spectacular flares. Massive amounts of water were needed to quench the flames. Just like steelmaking, coking used thousands of gallons of river water every day.

The water used in steelmaking tends to pick up whatever impurities that are rinsed off materials in the process. Some bright fellow had the idea of using dirty water from other parts of the mills to quench the coke, which made sense except that however poisonous the water was when it came from the mill, it would only be made worse by quenching. Mrs. LaMendola told me she could never get tomatoes to grow in the path where the plume from the ovens ran. On the other side of her house, they did just fine.

In the 1950s, the mills in Donora relied on blast furnaces each of which held 110 net tons of material. The basic process of steelmaking cooked iron ore into pig iron in these massive furnaces and then converted this into steel in open hearths. Molten iron solids would flow around the center of the blast furnace—an area where nothing moved, called the "dead man"—before dropping to the bottom as pig iron. Every few hours, about 1,450 tons of the heavier molten iron were tapped out into forty-ton brick-lined ladles that would be carried to the open hearths to become steel. Limestone served as a kind of chemical sponge, to dissolve silica and other impurities. When simmered to the right point, this amalgam floated above the liquid pig iron, where it could be skimmed off. Sluice gates from the furnace channeled the steaming leftover slag into waiting gondola cars shaped like giant teacups. The train was hauled to the dump, where its smoldering cargo was poured off between the surrounding bluffs, forming crater-like, jagged edged shapes as it cooled.

On summer evenings, my family and I would sit in lawn chairs behind our house and watch the fiery spray of what was called kish. Brightly burning graphite spewed off the ladles that drew hot iron from the furnace and burned like gigantic, brilliant sparklers. Each minute of the day all year long five vertical engines sucked in 42,000 cubic feet of hot air, yielding thousands of cubic feet of gas. When burned through a single stack atop the furnace, these gases plus lots of reddish iron ore and other dusts flamed at night like a rocket's trail. The spectacle was dazzling.

Making steel requires a furnace that stays constantly fiery to meld the metals. Into the shallow, saucer-shaped open hearths was placed a layer of scrap iron, some iron ore, coke, and limestone, onto which molten pig iron was poured. Preheated air mixed with gas forced the layers to melt into a white molten mass, at a temperature of at least 2,600 degrees Fahrenheit. At this heat, reactions in the cauldron forced out all air, leaving liquid metal that formed a dense, strong solid when it cooled.

The remaining white-hot liquid steel moved slowly on tracks inside the mill, where it was poured from huge ladles into five-ton molds to make ingots. An ingot was about the length of two men and the width of one—hot, heavy, and forbidding. After three hours, the ingots were stripped from their molds, still steaming red hot, and put into soaking pits before being shipped on open railroad tracks that connected the entire complex to the blooming and billet mills. Cranes mounted on massive beams built into the roof of the mill would pick up the ingots as if they were Tinkertoys and deposit them on a conveyer table. There they went through gigantic rolling pins for stretching and shaping into sheets, plates, and bars—the building blocks for the essentials of industrial life.

With much of steelmaking, there is little room for error. The carbon that remained in the final product determined whether what came out of the furnace was iron or steel. More carbon makes a harder, denser steel, but this steel becomes increasingly brittle.

A few years before I was born, a steelworker fell into the ladle used to draw off the molten brew, just as the furnace was being tapped. They said he'd been drinking, though how this was proved is beyond me. Not a single body part was recovered. They buried the bucket outside, near the furnace.

I loved the spectacular shower of sparks and sprawling fires that lit up the sky for miles, glowing with fiery dusts and gases from the furnaces. It was a fiercely hypnotic sight. My cousin Mark remembers that people on their way to Pittsburgh would stop their cars on the other side of the river just to watch.

The greatest enemy of steel, oddly enough, is air. Because oxygen is constantly trying to bind with iron to create the permanent orange layer better known as rust, steel had to be coated to keep out moist air. Products intended for outdoor use could be given a galvanic shield. This essentially meant plunging the steel into a bath of molten zinc at about 850 degrees. The zinc would bond to the surface of the steel, forming a series of layered zinc-iron alloys. When done properly, these alloy coatings last decades.

Before World War I, during the violent work slowdowns and protests that gave rise to the big steelworkers union, Donora remained staunchly antiunion. (In 1919, it would be the only town to oppose the Homestead strike.) For being the consummate company town in its early history, Donora was rewarded with a zinc plant.

The new zinc works built in 1915 was one of the world's largest facilities at the time; it stretched for forty acres along the river and was out of date the moment it opened. Its massive, horizontal coal-fired furnaces were already giving way to electrically powered plants that were less smoky and that did not create such quantities of toxic zinc fumes. The plant's smokestacks, moreover, were less than 150 feet tall, too short to propel their contents above the 600-foot hills around them. In 1933, after the plant had been firing for less than two decades, a Pennsylvania historian reported that bones from some old Native American graves had washed out of the hillside downwind of the zinc plant's plumes. The grass that held the earth in place had died off.

Working zinc was like coking, only worse. The zinc furnaces were so hot that you could see heat rising from them in rivers of distorted light, like fun-house mirrors. At its peak, the Donora Zinc Works employed about 1,500 men, who enjoyed an average workday of just three hours and yet received the highest wages in town—this in an era before unions had entered the plants. There was some difference of opinion about why this was. The workers themselves used to say it was because they were so efficient that they could fill the ovens in three hours with as much raw material as could be processed in an entire day. Lynn Snyder, an historian who studied the town's pollution, maintains that zinc workers worked a three-hour day because nobody could have tolerated more time than that in front of the red-hot furnaces.[1]

Most of the plant's employees had emigrated from parts of Spain where their families had produced zinc workers for generations. They did not mix much with the rest of the townsfolk. One fellow who had worked in a zinc plant in the 1950s commented to me, "I was the only one in the workforce who could read or speak English. Most of the workers were under twenty-five. Few of them lasted very long." He described his last day in the plant: "Five guys had gone before me to shovel out the finished zinc. Each one of them keeled over, real sick, kinda pale, and nearly passed out. I was the sixth one in. I couldn't take it either. I left. Spent a week in bed and never returned. Not many ever made it to the age of thirty as zinc workers. I quit when I was twenty."

Zinc is one of those elements that the body needs in very small doses in certain forms, but zinc can be poisonous in larger amounts and other forms. When bound with sugars in microdoses, zinc probably fights colds by killing rhinoviruses. When combined with gases of sulfur, carbon, fluoride, or nitrogen, zinc can be exceptionally dangerous. And it was not the only poison rising from those ovens. Zinc smelting and steelmaking both use lots of fluorspar, a rock made of crystals of fluorine tied with calcium. During smelting, fluorspar creates a penetrating and corrosively toxic fluoride gas that can eat the gloss off light bulbs, etch normal glass, and scar the teeth of children. One investigator found that mottled teeth, characteristic of fluoride poisoning, were common in Donora. My father had teeth like that. We figured he simply hadn't brushed enough as a kid.

Fumes from the mills, coke ovens, coal stoves, and zinc furnaces were often trapped in the valley by the surrounding hills. They gave us astonishingly beautiful sunsets and plenty of barren dirt fields and hills to play on.

On calm, cloudless, dry nights, the air gives up its heat to the surrounding hillsides, and growing denser as it cools, flows downhill like water. Usually, the temperature within any column of air is cooler the higher you get. Where there are valleys, the colder air from the hills can create an inversion layer that keeps warmer air from rising. Hot-air balloons fly because hot air is lighter than cold air. But when an inversion happens, balloons cannot fly, smoke cannot rise, and fumes, hot when released, cool and sink back to the ground, unable to dissipate.

October 26, 1948, brought a massive, still blanket of cold air over the entire Monongahela Valley. All the gases from Donora's mills, furnaces, and stoves were unable to rise above the hilltops and began to fill the homes and streets of the town with a blinding fog of coal, coke, and metal fumes. At first, cars and trucks tried to creep along with their headlights lit, but by midday, traffic came to a standstill as drivers could no longer see the street. "I could not even see my hand at the end of my arm," recalls Vince Graziano, then a strapping young steelworker. "I actually could not find my way home. I got lost that day."

Later, Berton Roueche, *The New Yorker*'s distinguished medical writer, described it this way:

> The fog closed over Donora on the morning of Tuesday, October 26th. The weather was raw, cloudy, and dead calm, and it stayed that way as the

> fog piled up all that day and the next. By Thursday, it had stiffened adhesively into a motionless clot of smoke. That afternoon, it was just possible to see across the street. Except for the stacks, the mills had vanished. The air began to have a sickening smell, almost a taste. . . . [2]

Arnold Hirsh, a World War II veteran then just beginning his half-century as the town's leading attorney, watched the gathering fog from his Main Street office: "The air looked yellow, never like that before. Nothing moved. I went over to Seventh Street and stood at the corner of McKean, looking down towards the river, and you could just barely see the railroad tracks. Right there on the tracks was a coal-burning engine puffing away. It issued a big blast of black smoke that went up about six feet in the air and stopped cold. It just hung there, with no place to go, in air that did not move."

The sturdy people of Donora were not perturbed. On Friday afternoon, the town's annual Halloween parade took place under a spooky haze. Children's costumes appeared and disappeared in the mist as the parade moved the two blocks down Main Street. My mother remembers it as a ghastly sight, but it fit the occasion. "Of course we all went," she told me later. "This fog was heavy, but there was only one Halloween every year. Only this time we could not see much." People could not see their own feet. Within days, nearly half the town would fall ill.

Donora did not abandon its routines easily. The high-school football team, the Dragons, practiced kickoffs on Friday in preparation for the next day's home game against their great rivals, the Monongahela Wildcats. At the practice, Jimmy Russell, the head coach, had to yell "Kick!" so that the receiving players would know the ball was in the air. He had no idea that some boys had taken advantage of the fog and left early.

The football game between Donora and Monongahela went off on Saturday as scheduled. The entire town turned out for pep rallies and parades, with strutting drum majorettes leading the black-and-orange-uniformed marching band. The spectators often lost sight of the ball and could only guess from the referees' whistles when to cheer. Donora's star tight end, Stanley Sawa, was ordered by the public address system in midgame to "Go home! Go home now!" Some in the stands thought it was a prank.

Still in his uniform, with his helmet in his hands, Sawa raced up and down the hills to his family's home at the bottom of Fifth Street, one of the many streets that were so steep they had stairs instead of sidewalks. He dashed into the house.

"What's going on?" he huffed. "Why'd you make me leave the game?"

"It's your dad," a neighbor told him.

"What are you talking about?" Sawa demanded. "Where is he?"

"In there, with the doctor," came the reply. "It doesn't look good."

The elder Sawa, who earned his living lifting massive loads of iron ore, had been brought home from the mill, short of breath, dizzy, thinking he only needed to lie down. By the time Stanley arrived home, his father had already died.

Monongahela won the game, 27 to 7. Spectators leaving the field later learned that by ten o'clock that Saturday morning, nine people had died. Within twenty-four hours the number would be up to eighteen.

Arnold Hirsh had tried to attend the game: "My brother Wallace and I decided we would walk up the Fifth Street steps. . . . We had just gotten out of the service. He had been a lieutenant in the navy, and I had been an infantry officer. We were both in as good shape as you could be. When we finally got to the top of those steps on our way to the game, we simply could not take another step. We did not say another word to each other. We could barely talk. We turned and headed straight home."

When they got there, they found their mother in distress. "My mother, who had not been well for years, just could not catch her breath," Hirsh recalled. Donora had eight doctors at the time, all of whom made regular house calls. This time, however, no one would come. "I called Doc Rongaus and he said that he just could not make it. He said, 'The whole town is sick. Even healthy fellas are dropping. Get your mother the hell out of town!'" The Hirshes drove into the Allegheny Mountains, away from the fog. Arnold's mother had come to Donora in 1920 as a healthy teenage bride. Both her parents, who lived elsewhere, survived to almost one hundred. By the time her two children were grown, she was an invalid with a weak heart and serious breathing difficulties. She died two years after the smog, having barely reached her fifties.

Doc Rongaus gave the same advice to anyone who would listen: Leave if you can. The firemen of Donora went from door to door delivering whiffs of oxygen from tanks to those who were stranded. One of the firemen, John Volk, remembered borrowing oxygen canisters from the Monongahela, Monessen, and Charleroi fire departments. "There never was such a fog. You couldn't see your hand in front of your face, day or night. Hell, even inside the station the air was blue. I drove on the left side of the street with my head out the window, steering by scraping the curb."

When I visited him in 1999 at an old age home, Doc Rongaus told me that folks who made it to Palmer Park seemed to recover. The park sat high on a hill and was one of the few green places near the town, probably because the fumes from the mills did not regularly sweep over it. "My brother and I hauled women and children in horse-drawn wagons up to the park," he said. "Soon as we got them above the smog, they would get much better." Church ladies from nearby towns provided food and blankets to the involuntary campers.

Others shut themselves in. "I had an elderly aunt and uncle," Hirsh recalled, "who lived on the corner of Fourth and McKean, named Myerson. My aunt looked out the window and figured out that this was something pretty bad. She closed her doors and kept them closed. They had no problem at all. They just stayed inside for five days."

The folks who ran the mills stuck to their routine. The whistles that kept the daily rhythms of the town shrieked on schedule, and the shifts that kept the plants running day and night did not cease. Although many people whispered that the mills had put something strange into the air, the superintendent of the zinc works, Michael Neale, knew that his mill was doing nothing unusual.

It was Walter Winchell, with the voice that resonated importance and certainty, who made Donora famous. "Good evening, America!" he said in his national radio broadcast that Saturday night. "The small, hard-working steel town of Donora, Pennsylvania, is in mourning tonight, as they recover from a catastrophe. People dropped dead from a thick killer fog that sickened much of the town. Folks are investigating what has hit the area." But he had already given the answer many would come to accept: It was a "killer fog," a freak of the weather—ultimately, an act of G-d.

That weekend, the enormous volume of telephone calls created a five-hour wait before frantic relatives could speak to local residents. Roger Blough, then chief counsel of American Steel and Wire and later its CEO, finally reached Neale at three o'clock Sunday morning to tell him to dead-fire the furnaces, without zinc ore. A zinc furnace, like a coke oven, cannot be allowed to stop; once cooled, it can never be restarted. Dead firing would protect the equipment while reducing the plant's emissions. Resentful of the interference and unconvinced that there was a problem, Neale only complied after a group of company-hired chemists arrived at six that morning, three hours after he had received the order to reduce operations. He described this action as a gesture of concern for the community, not an admission of responsibility for the smog. As he later told the press, "the zinc works has operated for thirty-two years with no problem."

By the time the fog began to ebb that Sunday, the local funeral home had run out of caskets. The basement of the community center, where the Brownies, Cub Scouts, Girl Scouts, and Boy Scouts usually met, became a temporary morgue. The *Pittsburgh Post-Gazette* reported, "the citizenry maintained an attitude of outward calm which was surprising to observers. . . . Here and there on the streets the youngsters continued their games of touch football and rode their bikes." Rains early on the morning of November 1 washed the skies of whatever had hit the town. By November 2, the zinc mill was again running at full steam. The same work ethic that kept the football team practicing, the marching band playing, and the cheerleaders cheering also got the town right back to work.

Roueche wrote,

> Funeral services for most of the victims were held on Tuesday, November 2nd. Monday had been a day of battering rain, but the weather cleared in the night, and Tuesday was fine. "It was like a day in spring," [funeral director] Schwerha says. "I think I have never seen such a beautiful blue sky or such a shining sun or such pretty white clouds. Even the trees in the cemetery seemed to have color. I kept looking up all day."[3]

The day after the funerals, Joseph Shilen, a county medical official, filed a report with the Pennsylvania secretary of health, recommending that the zinc works be reopened. The incident, he wrote, was unlikely to recur. Asked to investigate the smog, John J. Blumfield, deputy head of industrial hygiene for the Public Health Service, refused to do so, calling it "a one-time atmospheric freak."[4]

What happened in Donora was not freakish, nor was it the first time that winds and weather combined with industrial fumes to kill so many that the deaths could hardly be counted. Neither Donorans nor many others knew that in 1930, in the Meuse River valley of Belgium, dozens of people had died within days in a smoky fog. Here, too, the exact count was never tallied. Like Donora, the city of Liège sat on a series of steep hills around a river valley, surrounded by metal mills and smelters. The conditions were similar: heavy fogs, lots of fumes from the mills, and workers who depended absolutely on the mills to feed their families.[5]

Some experts who studied the 1930 Liège disaster six years later warned of the consequences if a similar catastrophe were to befall a larger city. Given the size and age distribution of Liège's population, they calculated that if the same conditions ever hit London, more than four thousand people would die in a single week. Nobody listened.[6]

One Belgian investigator painstakingly demonstrated that fluoride gases were the likely cause of the devastation. Sulfur, he pointed out, in heavy doses leaves distinct marks on the linings of the lungs, but fluoride gases do not. They pass right into the bloodstream and attack the heart and other organs, without marring the nasal passages, throat, or lungs. The lungs of those who died in Liège were clean. Nobody noticed.

The work of epidemiology is easiest whenever exposures are deliberate and controlled and take place over relatively short periods. The gold standard for epidemiologic research is the randomized, controlled trial of pharmaceuticals or surgical techniques. Under this system, people with similar personal characteristics, age, weight, height, medical histories, and so forth, are assigned to be given the test intervention or not, and their progress is carefully charted and tracked. Those who test drugs in this fashion have the luxury of studying controllable, easily identifiable causes intended to produce known effects. They can set up studies that ensure that a given result reliably comes from the intended source.

The world in which we live and work is not such an easy or elegant place to study. When it comes to studying the impact of the environment on health, epidemiology remains an inexact instrument. In contrast to the precisely metered world of clinical trials of drugs, the environment in which people actually live is complex, unyielding of its secrets, and generally uncooperative. People outside of these controlled trials seldom know what they've been exposed to even recently, let alone in the critical weeks just before and after their birth. Getting accurate long-term-exposure data is even harder.

Most often, we fail to find solid answers. It took fifty years of finding unmistakably higher levels of sickness and early death in smokers for us to reach the conclusion that cigarettes really are bad for you. Our track record on environmental effects on health is especially unimpressive. As one of my colleagues remarked, "an epidemic is something so obvious it can be detected even by epidemiologists."

There are many reasons for these failures. What we do is genuinely difficult. Epidemics, which are by definition brief and localized, sometimes end before anybody can come in to study them. Sometimes we don't ask the right questions, or we ask them in such a way that the answer remains unobtainable. Even worse, sometimes the right questions are asked and even answered, but the news remains locked away in someone's private files or is written so abstrusely, and published so obscurely, that it might as well not exist at all.

For Donora, as for Liège, the important questions never got asked. Information critical to figuring out what went on remained hidden, sometimes in full view. As a result, the right things never got counted.

After the smog, a brief campaign erupted against the zinc mill, led by folks outside of Donora. Abe Celapino, a prosperous farmer and restaurant owner from across the bridge in Webster, whose cows and chickens had died, joined forces with the *Monessen Daily Independent* in calling for the mill to be relocated to a desert area. The editor in chief remarked that this might soon be unnecessary: The mill was creating its own desert area where it stood. Dr. Bill Rongaus, then the only member of the Donora Borough Council who was not employed by the mills, pointed out that the zinc mill was likely to account for the sudden sickness. "There was fog in Monessen, too," he told the Donora Board of Health, "but it didn't kill people there the way this did. There's something in the air here that isn't found anywhere else." Celapino alleged that Michael Duda, a zinc worker and borough council member, had told him late one night in Celapino's restaurant, "I've got a darn good job and I'm going to keep it. I don't care what it kills."[7]

In the month that followed, calls for major studies of the town were rebuffed by people who did not want to know the answer and by others who feared what it would ultimately mean for the town's workforce. It was revealed that the town council and the Chamber of Commerce had requested advice from the Pennsylvania Department of Forests and Waters the previous March. A reply from Deputy Forester James Cornely was read to a community hearing just after the smog had cleared: "It is my belief that Donora could demand that smoke filters could be placed in the smoke stacks of the zinc plant; and if done in the right manner with the suggestions of a possible usable precipitate or residue being produced, the result might be satisfactory."

This was an early suggestion for what later became standard industry practice. The escaping fumes contained valuable metals and other materials that could be trapped and reused, netting the mills more money and the town less pollution. But the mill operators in Donora had no interest in such a device.

The steelworkers union, not realizing it was putting jobs at risk, offered $10,000 for a study that would explain the sudden deaths. The study, though delayed for two months and begun only after the mill had switched from coal to natural gas in some critical departments, was suitably ambitious. The first folks in were the medical experts. In a massive effort, nurses armed with questionnaires surveyed half the homes in Donora. Then the pathologists mounted an intensive study of the twenty who had died during the fog itself. The doctors conducted all the usual clinical tests. They looked into each lung, each heart, and every other tissue that could be stained and assembled. A preliminary report by the Public Health Service, full of details of blood tests and other procedures and illustrated with copious x-rays and slides of lung tissue, came out within a year, but no final report followed. The detailed medical histories the nurses gathered have never been found.[8]

One eyewitness report from a medical expert paints a convincing picture of fluoride gas poisoning. "Listening to the affected chest, nothing could be heard. Occasionally inspiratory and expiratory wheezes would be heard in the asthmatics, but in the healthy chest nothing at all. It was as though the respiratory apparatus was paralyzed. Many were cyanotic [blue] and apprehensive not knowing what had happened." This expert did not identify a cause, but he could easily have been describing the clean lungs of the victims in Liège.[9]

A Philadelphia chemist brought in to study the problem, Philip Sadtler, speculated that the toxin came directly from the mills. Within months of the disaster, he reported that he had found over 1,000 parts per million of fluoride in an air-conditioning unit from Donora. Blood taken from those who died showed twelve to twenty-five times the normal levels of fluoride.[10]

Their lungs, Doc Rongaus recalled some fifty years later, often looked fine at autopsy. A report issued by the commonwealth of Pennsylvania corroborates his memory, especially in its description of the person identified as Case P:

> The evidence . . . discloses that the larynx, trachea and bronchi of the first order were little affected. Apparently, the irritating agent was carried into the lung and exerted its primary effect upon the terminal bronchi, the bronchioles and the pulmonary parenchyma. . . . However, the agent must have had a low irritating capacity since none of the cases exhibited a degree of hemorrhage, oedema, or necrotizing process commonly associated with the inhalation of lethal irritating substances. Analogy might be made here with certain war gases. Phosgene, for example, has little effect upon the upper respiratory tract. The finer bronchi and lungs undergo intense oedema and congestion during the acute phase of the poisoning.[11]

In other words, the body's upper breathing system was not disturbed by the air in Donora. Whatever killed these people slipped deeply and directly into the body, making a bloody swollen mess of the lower lungs, much like phosgene, a nerve gas used in World Wars I and II.

The source of the poison in Donora was never identified. The lethal smog spawned an entire new academic profession, focusing on the study of humans exposed to polluted air. The Public Health Service was charged with analyzing, assessing, measuring, and confirming what had happened. Donora was investigated to death—not because so many studies were done but because the absence of definitive evidence of air pollution's harm was taken as proof of its safety.

The few investigators who warned that all this was not merely bad weather were dismissed. About a year after the inconclusive Public Health Service report was issued, a remarkably candid critique of the report appeared in *Science* magazine on January 20, 1950. Clarence A. Mills, a physician from the University of Cincinnati, had been trying for some years to generate support for studying the conditions of the Monongahela Valley. He wrote that just two years before the disaster, there had been no interest in such research. Now he asked: "Just what did their year's work, with a staff of 25 investigators show?"[12]

The answer was, pitifully little:

> The most valuable part of their year's work—analysis of poison output from the steel and zinc plant stacks—remains unused and unevaluated in their written reports. They spent months analyzing the valley air for poisons, but failed to calculate the concentrations probably present during the killing smog a year ago, when an inversion blanket clamped a lid down over the valley's unfortunate people. Had they made such calculation, they would have found that even one day's accumulation of the very irritating red oxides of nitrogen from the acid plant stacks would have caused concentrations almost as high as had been set as the maximum allowable for safety of factory workers exposed only for an 8-hour work day. At the end of 4 days of last year's blanketing smog, concentrations reached were probably more than four times higher than the 10 milligrams per cubic meter of air listed as the upper limits of safety! And

> the Donora people breathed the poisoned air not 8 hours a day but for 4 whole days.[13]

Mills noted the eerie and tragic parallels between Donora and Liège, where nearly identical conditions created lethal brews involving low-lying mill towns and zinc and steel fumes. And he challenged the claim that the Public Health Service had opened up a new field of inquiry, charging the organization with ignoring years of work by others:

> Let us hope that the Donora tragedy may prove such an object lesson in air pollution dangers that no industrial plant will feel safe in the future in pouring aloft dangerous amounts of poisonous materials. Let us hope that the Donora disaster will awaken people everywhere to the dangers they face from pollution of the air they must breathe to live. These 20 suffered only briefly, but many of the 6000 made ill that night will face continuing difficulties in breathing for the remainder of their lives. Herein lies the greater health danger from polluted air—continuing damage to the respiratory system through years of nonkilling exposure.[14]

It has a strange ring to it, "years of nonkilling exposure."

The Public Health Service did not agree that the zinc works played any important part in the deaths. Despite independent tests showing that even sixty days later, air concentrations of fluoride gas were ten times what was then considered safe, the health service team made no measurements of fluoride levels for itself and did not mention the possibility of fluoride poisoning in its preliminary report.

In Donora, efforts to link the disaster to fluoride fumes and other metal fumes were fiercely contested. The Pennsylvania Department of Forests and Waters had complaints on file from farmers downwind of the zinc mill dating back to 1915, but the Chamber of Commerce, according to Doc Rongaus, saw that these reports never surfaced. Michael Neale, as leader of the Chamber of Commerce and head of the plant that emitted more fluoride gases than any other, was apparently determined not to endanger his production goals just because some folks had gotten sick.

And Donora seemed to side with Neale. Even as citizens from surrounding areas urged that the smog serve as the impetus for cleaning things up, no serious support for this position arose within the town itself.

I asked Rena Hirsh, Arnold's wife of half a century and a town resident for all of their married life, what she knew about the studies in Donora on people who had been "non-killed." "You'll have better luck finding a needle in a haystack than finding records of what really went on," she replied. "I don't know for sure. I might be wrong. But people really did not want to know what had happened."

It is possible to look without seeing. Once I went on a tented safari between Kruger National Park, Timbavati, and the Sabi Sand Reserve in Manyeleti, in Northeast South Africa. I sat in an open vehicle some twenty feet from a large clump of dry, beige bush on the savanna. The tracker whispered, "Look, there's the lion!"

I stared hard but could not see a thing. "He's right there!" the tracker insisted. Still I could see nothing. I thought the guide was playing with me. In front of me tan grass ran in all directions, broken only by occasionally larger clumps.

In the instant before the lion roared and charged at a speed I had not imagined possible, I managed to make out one dark eye. The cat pulled up about four feet from our jeep and slowly sauntered away. Lions, we were told, often bluff.

Looking without seeing is something others do. It took a charging lion to convince me that I might do it too—look at data over and over and not see what was right in front of me.

The first medical experts into Donora after the smog conducted all the proper clinical tests on the twenty who died right away. Following traditional approaches, they looked into each of the vital organs and all the other tissues that could be stained and assembled. They looked at each slide, each x-ray one at a time, and never put them all together. No one measured pollution in Donora until two months after the fatal smog had ended. As Mills noted, no effort was made to reconstruct what had gone on during the episode itself. Worse, the experts never looked at the survivors. If they had, they would have learned that in the month after the smog lifted, at least fifty extra people had died.

The notion of "extra" deaths may seem strange. As my mother says, you only get one chance to die. But epidemiologists can, and routinely do, predict the number of people who should die in any given population in any given period, and thus can tell if a group of deaths is occurring that should not. These statistical patterns of dying are human lives with the tears removed, the literal bodies of evidence. In Donora, one of every three people got very sick during the week-long smog. Even a decade later, the town's death rate was much higher than in surrounding towns. But no attempt was made to link these deaths to the smog or to air quality in general.

The few results that did emerge were all published in the Public Health Service's preliminary (and, as it turned out, final) report of 1949. This report, which Mills found so lacking, now sits in the offices of the few people in the world who are concerned about the matter.

One of these is Robert K. Maynard, the adept and irreverent head of the Environment Program in central London. His idea of fun is to amass historical documents on interesting cases of pollution crises and their health effects. When he heard about my research on Donora, Bob showed me his collection of original maps of the area, lung slides, and other documents. "Here, you might find these amusing," he said, handing me a foot-tall stack of old reports and newspaper articles. "Let me know what you find out. It might come in handy. I could send you some photographs too if they'd be useful." I dreaded having to carry the stuff, but I did, and I accepted Bob's offer to send more. About a month later, a heavy package arrived, full of pathology reports, autopsies, and lots of official documents.

For two years, I sifted through these papers, over and over. When I finally found the key to Donora, it was something I had looked at countless times but never seen.

A simple black-and-white map shows the homes of each of the eighteen people who died in Donora between October 26 and 29, 1948, and the two who died shortly after. I transferred each death to a larger map that showed the hills and valleys, the streets and the mills.

These records had been sitting in various files for half a century. Plenty of people had turned their microscopes on each one of them, but nobody had stepped back to see the pattern.

It was not merely a sudden bad break of weather. It was foggy then, but the valley is still foggy in the fall today, and the fogs will continue for as long as warm river water emits vapor into colder air. What killed the people in Donora was what many suspected but could never prove. Most of the deaths occurred in the parts of town that sat just under the plume that spewed within a half-mile circle of the zinc mill.

Many events of the 1940s—Pearl Harbor, D-Day, the end of World War II—were commemorated with fiftieth-anniversary celebrations. The fiftieth anniversary of Donora's killer smog almost passed without notice.

Donora is a different place now. After the big strikes in the fifties, the massive, inefficient mills shut down, leaving the town to cope with deteriorating schools and a crumbling tax base. Many of the men, unwilling to give up the homes they had so painstakingly built (or unable to sell them) began commuting sixty or even a hundred miles a day to take jobs in other towns. The Monongahela, ever a poisonous brown, began to flow blue. "First they tore down the big plants," one resident recalled. "Then they built a McDonald's and no one came. So they boarded that up and built a parking lot. Now nobody parks there either." Main Street now has a single traffic light, a second one having been converted back to a four-way stop sign to save on maintenance. The former Hotel Donora, once home to dozens of bachelor mill workers and the town library, is a martial arts training center and occasional rooming house.

On one of those lovely spring days when the meandering Monongahela basin looks impossibly green, Arnold and Rina Hirsh walked with me up to the small, neatly kept Jewish cemetery where my grandparents are buried. It sits atop one of the prettiest bluffs in the area, with a panoramic view of the rolling river valley. The Hirshes, Sammy Baylis, and Herman Weiss keep the place locked and tidy, with small gray headstones recalling Donora's former residents and waiting for the few who remain to join them. "You see the beautiful trees and beautiful view we have today?" Arnold said to me. "There was nothing here for many years after the smog was over. There was nothing but clay on that hillside."

In 1998, just about the time of the fiftieth anniversary of the disaster, an earnest high-school student named Justin Shawley got a monument set up. The Pennsylvania Historical and Museum Commission erected a five-foot-square bronze plaque near the center of the former steel mill as a memorial to those who died. To mark the occasion, residents and local and state officials held a service at Our Lady of the Valley Catholic Church, one of Donora's few remaining houses of worship. The plaque says:

The 1948 Donora Smog

> Major Federal clean air laws became a legacy of this environmental disaster that focused national attention on air pollution. In late October of 1948, a heavy fog blanketed this valley, and as the days passed, the fog became a thick, acrid smog that left about 20 people dead and thousands ill. Not until October 31 did the Donora Zinc Works shut down its furnaces—just hours before rain finally dispersed the smog.

It is a touching monument. The fifty people who died in the month following the smog are nowhere counted. The thousands who died over the following decade are nowhere counted. And there is no counting of the thousands whom Clarence Mills called the non-killed—all those who went on to suffer in various poorly understood ways. Standing there by the ruins of the old mill, I thought I understood, just a little, what Sol Filler must have felt on revisiting Tereisenstadt: These people are well intentioned. They are trying to commemorate, to remember, to atone. But they are not trying hard enough.

Every one of Bubbe Pearl's five children developed heart problems. None of their illnesses would ever be tied to where they grew up. They are not listed on any memorial to Donora's dead. In 1969, my dazzling, athletic Uncle Len dropped dead at age fifty on a handball court in Southern California, years and miles away from the Monongahela River Valley. But he carried Donora with him in his heart—and in other body tissues as well. By the time my mother reached the same age, a decade later, coronary artery bypass operations were available to keep her alive. She needed three of them. Aunt Gert required only two angioplasties.

Bubbe Pearl's tombstone sits in the lovely Jewish cemetery with its spectacular view of the river valley. When I was born, she was still a fierce driver, but by the time my brother arrived, a year and a half later, she had become an invalid. She did not die during the smog of 1948 either—the town erected no plaque for her—but only some two dozen heart attacks later. The attacks were so common that they became almost a ritual. The room would go quiet, and my mother, the baby of the family, would steady her own mother by the arm and steer her to the bed. The heavy, mottled-green, steel oxygen tank would be wheeled over, the valve turned on, and the gas mask pulled over Bubbe's nose and mouth. Her skin often matched the blue-white color of her hair.

Aunt Gert, the oldest sister, always had to leave the room, unable to stand by helplessly while her mother fought for air. Sometimes Bubbe would shriek, "*Oy vey!*" But usually there was silence, and sighing. We would all wait for Dr. Levin. Dr. Levin always came, always calm, always sure. His arrival meant that everything would be all right.

The night she finally died, he did not come. I could not stop crying. I had seen her nearly die so many times, I was sure it was a mistake.

2

THE PHANTOM EPIDEMIC

It was a foggy day in London, and the fog was heavy and dark. Animate London, with smarting eyes and irritated lungs, was blinking, wheezing, and choking; inanimate London was a sooty spectre, divided in purpose between being visible and invisible, and so being wholly neither. Gaslights flared in the shops with a haggard and unblest air, as knowing themselves to be night-creatures that had no business under the sun; while the sun itself, when it was for a few moments dimly indicated through circling eddies of fog, showed as if it had gone out, and were collapsing flat and cold. Even in the surrounding country it was a foggy day, but there the fog was grey, whereas in London it was, at about the boundary line, dark yellow, and a little within it brown, and then browner, and then browner, until at the heart of the City—which call Saint Mary Axe—it was rusty-black. From any point of the high ridge of land northward, it might have been discerned that the loftiest buildings made an occasional struggle to get their heads above the foggy sea, and especially that the great dome of Saint Paul's seemed to die hard; but that was not perceivable at their feet, where the whole metropolis was a heap of vapour charged with the muffled sound of wheels and enfolding a gigantic catarrh.

—Charles Dickens, 1865
Our Mutual Friend

Among the ways you can amuse yourself as a tourist in England is to wend your way through the narrow streets of East London following the trail of Jack the Ripper, the notorious gaslight-era killer of young women. You can join a group of fifteen or twenty others, huddled under one of the Thames bridges waiting for the drizzle to slow, to be herded along by the lilting, demandingly impeccable tones of an English tour guide.

It's better if you go, as I did, at night. The lumbering visit to the dark side of London takes you past the Ripper Museum, with its neon signs boasting plastic reenactments and a few authentic remnants of grisly crimes, and right by the Tower of London and the original city dungeons. More than a century later, the area is still pretty grimy—not a place women should venture alone. Perpetually wet cobblestone streets still run alongside the river. Fogs still roll in as the daytime temperatures drop. But the mists no longer bring the city to a standstill.

At the end of the nineteenth century, wet air would rise from the river, mix with the dark coal smoke, and hang in the streets like chilled fudge. These were no ordinary mists but an intense, stifling smoke in which a woman could not see her own feet. Traffic all over London came to a halt, and even horse-drawn ambulances stopped in their tracks. It was under the cover of such fogs, from August to November 1888, that the Ripper waited in the dim, cobbled alleys for the kind of women who walked these streets at night. Jack the Ripper was never caught, but we know the identity of his accomplice. He escaped into the clotted London fogs that fascinated writers and artists for centuries.[1]

Charles Dickens, the chronicler of nineteenth-century manners and morals, began his novel *Bleak House* in 1853, with a passage permeated by fog:

> Fog everywhere. Fog up the river, where it flows among green meadows; fog down the river, where it rolls defiled among the tiers of shipping, and the waterside pollutions of a great (and dirty) city. Fog on the Essex marshes, fog on the Kentish heights. Fog creeping into the cabooses of collier brigs; fog lying out on the yards, and hovering in the rigging of great ships; fog drooping on the gunwhales of barges and small boats. Fog in the eyes and throats of ancient Greenwich pensioners, wheezing by the firesides of their wards; fog in the stem and bowl of the afternoon pipe of the wrathful skipper, down in his close cabin; fog cruelly pinching the toes and fingers of his shivering little 'prentice boy on deck. Chance people on the bridges peeping over parapets into the nether sky of fog, with fog all around them, as if they were up in a balloon, and hanging in the misty clouds.[2]

Heavy-metal pollution, collapsed sunlight, and heaps of vapor captivated not only the writers of the day; the hypnotically vibrant red and blue sunsets also inspired artists. The impressionist painter Claude Monet preferred the ever-shifting light of the dark city to the more ordinary, bright countryside. In London's fogs he saw

> all sorts of colours . . . black, brown, yellow, green, purple fogs, and the interest in painting is to get the objects as seen through all these fogs. My practiced eye has found that objects change in appearance in a London fog more and quicker than in any other atmosphere, and the difficulty is to get every change down on canvas.

David Bates, the renowned medical essayist, has pointed out that this foggy blurring of details inspired more than one hundred of Monet's paintings of London's locales.[3]

London's reputation for coal-smudged skies did not arise overnight. For centuries, the city had the world's largest concentration of coal stoves, most inhospitable airs, and regularly foggy weather. In a fascinating essay on air pollution in London since medieval times, historian Peter Brimblecombe explains that in the Middle Ages, mountains of coal piled up in the city as a consequence of sea trade. From its large port, London regularly sent out vessels laden with animal hides, whale oil, tallow, dried fish and meats, fertilizer, and wools. Ships often came back from the less populated northern British Isles empty, save for the crew. To weather the rough seas around the coast, mariners filled their holds with what became known as sea-coal, *carbonem marus.* Vessels dropped off huge mounds of this worthless rock, hauled from Scotland and Ireland, so that they could return full of England's finest goods.

At first the coal was used solely as ballast, to make the empty ships more stable. But by the thirteenth century, mounds of dusty, black rock clogged city streets, docks, and alleys. There was a Sacoles Lane in London as early as 1228. Today both Seacoal Lane and Old Seacoal Lane are found near Ludgate Circus, an early center of fertilizer manufacture. Brimblecombe explains that one of the earliest industrial uses of coal fires was to convert limestone to lime fertilizer. When mixed with water, limestone also formed cement for building.[4]

In an effort to rid themselves of the black heaps that filled the city, Londoners in the thirteenth century began to use coal for many other purposes. It soon became apparent that coal burned longer and hotter than wood. Not all were enthralled with its smoke, however, even from the earliest days. Queen Eleanor, fleeing the fumes created by heavy use of sea-coal in Nottingham Castle in 1257, issued one of many fruitless royal bans on coal burning. In the last quarter of the thirteenth century, coal fumes created so many problems that a royal commission was convened to investigate the city's foul smokes and smells. Another royal proclamation banned the use of sea-coal in 1306, warning that those who deviated would be "punished by *grievous ransoms*."[5] None of these efforts stifled coal burning among foundries, lime burners, and other small firms. Practical needs prevailed.

By the fourteenth century, people regularly took coal into their homes for heating and cooking. As urban centers expanded in the region of greater London, the forests began to recede. Once trees could no longer be easily cut down within easy reach of the expanding city limits and wood for cooking and heating was not in ready supply, dependence on coal was set. By the fifteenth century, London's skies were regularly blackened with coal smoke.

Churchmen who held government posts in those days were accorded considerable leeway. The Church of England's Archbishop of Canterbury, William Laud, who served in the treasury as well as the church, fined the coal-fired brewers of Westminster to get funds to repair smoke damage to Saint Paul's Cathedral. This action, together with a habit of exacting double taxes from those who resisted the church's authority, was among the many reasons Laud was beheaded in 1645.[6]

By the seventeenth century, the practice of burning coal indoors had become a well-established nuisance. In a 1604 pamphlet entitled *Counterblaste to Tobacco,* King James I warned against "the vile custom of tobacco taking" and "coal-besotted kitchens."[7] He reasoned that tobacco smoking should be banned and coal smoke discouraged because smoke itself was vile and led to the intake of foul matter. "The smoky vapors sucked by the sun and stayed in the lowest and cold region of the air are contracted into clouds and turned into rain and such other watery meteors. So this stinking smoke being sucked up by the nose and imprisoned in the cold and moist brains is by their cold and wet faculty turned and cast forth again in watery distillations, and so are you made free and purged of nothing."[8]

Undeterred by the fate of the archbishop and the failure of royal decrees, the renowned English diarist John Evelyn, one of the founding members of the Royal Society, added fuel to the campaign against coal smoke. In 1661, he wrote what he called an "invective on coal burning" in a classic work, *Fumifugium, or the Inconvenience of the Aer and the Smoak of London Dissipated.* Scandalized that heavy smoke from Scotland Yard's coal stoves kept him from viewing the palace of King Charles II at Whitehall, and convinced that it was unhealthy for people and for the trees he so adored, he set his sights on ridding the city of the source of the scourge:

> Columns and Clouds of Smoake, which are belched forth from the sooty Throates [of small industries] . . . rendering [London] like the approaches of Mount-Hecla. That hellish and dismal cloud of sea coal [means] that the inhabitants breathe nothing but an impure and thick mist, accompanied by a fuliginous and filthy vapour, which renders them obnoxious to a thousand inconveniences, corrupting the lungs and disordering the entire habit of their bodies, so that cattarrhs, phthisicks, coughs and consumption rage more in that one City than the whole Earth besides. . . . Is there under Heaven such coughing and snuffling to be heard as in the London churches where the barking and spitting is incessant and importunate?[9]

Newcomers to London were especially hard hit:

> Those who repair to London, no sooner enter into it, but they find a universal alteration in their Bodies, which are either dryed up or enflam'd, the hunours being exasperated and made apt to putrifie, their sensories and perspirations so exceedingly stopp'd, with the losse of Appetite, and a kind of general stupefaction, succeeded with such Cathars and Distillations, as do never, or very rarely quit them, without some further Symptomes of dangerous Inconveniency so long as they abide in the place; which yet are immediately restored to their former habit, so soon as they are retired to their Homes and enjoy the fresh Aer again.[10]

Evelyn envisioned a city whose inhabitants would not have to struggle for air or pick their way among heaps of rubbish, dung, and other leavings of urban life. *Fumifugium* included several simple ideas, such as planting flowers, establishing parks, and banishing the smoky trades to the outskirts of the city.

Evelyn saw nothing impractical about this plan. His remedy, he wrote, required only

> the Removal of such Trades, as are manifest Nuisances to the City, which, I would have placed at farther distances; especially, such as in their Works and Fournaces use great quantities of Sea-Cole, the sole and only cause of those prodigious Clouds of Smoake, which so universally and so fatally infest the Aer, and would in no City of Europe be permitted, where Men had either respect to Health or Ornament. Such we named to be Brewers, Diers, Sope and Salt-boylers, Lime-burners, and the like: These I affirm, together with some few others of the same Classe removed at competent distance, would produce so considerable (though but partial) a Cure, as Men would even be found to breath a new life as it were, as well as London appear a new City, delivered from that, which alone renders it one of the most pernicious and insupportable abodes in the World, as subjecting her Inhabitants to so infamous an Aer, otherwise sweet and very healthful.[11]

At first, Charles II, an avid reader and a personal acquaintance of Evelyn's, appeared to embrace these suggestions. As wind of the king's inclinations got out, one can only imagine the response of councilors and the local tradesmen to the notion that they would have to give up their prime spots right near rivers and trade centers. Because it appealed to the British delight in blooming plants and formal parks, Evelyn's call for herbal gardens and flower beds survived. Saint James had been a low-lying swampland until 1660, when the royals cleared it out, planting a formal English garden that remains unrivaled. The royal family still maintains some of the loveliest parks in the city of London, including Saint James, Regent's Park, and Kensington Gardens. But while London's parks took root, many of Evelyn's other ideas did not. The smoky firms remained in place, even when a tragic opportunity arose to move them.

The Great London Fire of 1666 started on September 2, from an oven fire in the Royal bakery that had not been properly doused. Sometime after midnight, the fire crossed Fish Street Hill in central London. Sparks spread rapidly through the town's center, blown by the night's strong winds across the straw-thatched roofs and pitch-covered buildings. After a dry summer, the buildings ignited like kindling. From the Star Inn, the fire took down Saint Margaret's Church and then passed to Thames Street. By dawn the flames were halfway across old London Bridge.

From the vantage point of his home outside the fire, Evelyn described a living hell:

> Oh the miserable and calamitous spectacle! . . . All the sky was of a fiery aspect, like the top of a burning oven, and the light seen above 40 miles round about for many nights.
>
> God grant mine eyes may never behold the like, who now saw above 10,000 houses all in one flame; the noise and cracking and thunder of people, the fall of towers, houses, and churches, was like an hideous storm, and the air all about so hot and inflamed that at last one was not able to approach it, so that they were forced to stand still and let the flames burn on, which they did for near two miles in length and one in breadth. The clouds also of smoke were dismal and reached upon computation near 50 miles in length.[12]

Growing fires spew out sparks that can travel a half mile. Depending on where they land, these embers can die or fuel new flames. In addition to its many straw-thatched, wooden structures, the storehouses next to the river contained oil, tallow, brandy, and other combustible goods. The volunteer brigade was overwhelmed. The only way to stop such a blaze, now as then, was to create a firebreak by removing all fuel. In this case, that meant destroying every building in the fire's path.

The diarist Samuel Pepys lived in a house that would remain just outside the ring of fire. He knew immediately that this was no ordinary blaze and made a furious effort to direct the fight. Aware that the local officials would not take responsibility for destroying existing buildings, he went directly to the king, who gave the order to create a firebreak. Racing around the edge of the fire, Pepys brought the king's personal message to Lord Mayor Bludworth to take down houses and buildings. His diary records the ineffectual response:

> At last met my Lord Mayor in Cannon Street, like a man spent, with a handkerchief about his neck. To the King's message he cried, like a fainting woman, "Lord, what can I do? I am spent: people will not obey me. I have been pulling down houses, but the fire overtakes us faster than we can do it." . . . So he left me, and I him, and walked home; seeing people all distracted, and no manner of means used to quench the fire. The houses, too, so very thick thereabouts, and full of matter for burning, as pitch and tar, in Thames Street; and warehouses of oil and wines and brandy and other things.[13]

By this point, with the fire racing beyond control, there was no alternative. Buildings had to be destroyed to rob the fire of its fuel. Of course, Mayor Bludworth was concerned as to who would foot the bill for rebuilding. But Pepys's direct orders from the king forced the matter. "Trained bands" were called to assist with the demolition. At first, they could not get ahead of the inferno.

In desperation, people used explosives to blow up houses—often with excessive success. For three more days, the fire raged through the city. As relief began to set in, the dying fire flared up again and crept onward toward Whitehall. The Duke of York ordered the destruction of still more buildings. The fire finally burned out at Temple Church, near Holborn Bridge.

The fire lasted five days. A 1½-by-½-mile area lay in ashes, 373 acres inside the city walls and 63 acres outside. Eighty-seven churches, including Saint Paul's Cathedral, and 13,200 houses were destroyed. Only six people are known to have been killed, although the true toll was probably much higher. Still, the fire may have saved many more than it killed. Most of the rats that had helped to transmit the bubonic plague the previous year had died in the fire. The number of plague victims dropped rapidly afterward.[14]

As the Great Fire had destroyed nearly four-fifths of the old town, it set the stage for the replacement of wood structures with stone, the growth of insurance companies, the start of modern urban fire fighting, and an incidental rodent-control program. All this was welcome. Evelyn, however, had one other idea for London at that time—an idea that could have saved millions of lives.

With a nearly blank slate, the city could have been redesigned. Evelyn had urged that the major producers of smoke be located in a common industrial zone, far from where most people lived.[15] But he failed to reckon with the traditions and convenience that kept these same businesses as they were. Despite his friendships with those in power, Evelyn was no match for the economic forces of the day. In the seventeenth century, as in the twenty-first, pressures to keep things as they have always been could be far more powerful than well-founded suggestions for improvement.

Evelyn did not lack passion in his views about the dangers of smoke. He offered cogent reasons why he believed fumes to be unhealthy. But he could not marshal any sort of statistical, scientific, or empirical argument showing that such airs truly harmed living creatures. Seventeenth-century London did not lack people who could appreciate such an argument, had it been presented to them, but the intellectual tools for mounting it did not yet exist.

The capacity of people to get on with what they have been doing all their lives, even when they know it is not in their best interests, is a marvel. Denial is one of the strongest of human emotions. It gets us through the shock of chronic illnesses or sudden deaths, and often it is what keeps us from making changes in life. Thus it is not enough to have a good idea or even a great one to bring about social change. People have to believe that the problem being addressed is so bad that something must be done, and they must believe that something can be done. For the ideas of John Evelyn to take hold would take three more centuries and thousands and thousands of unnecessary deaths and illnesses.

A contemporary of Evelyn's, the self-made scientist-businessman John Graunt, created the tools that eventually allowed people to understand just how smoke and fires and other components of the world around us affect health. A prosperous merchant and art collector who lost everything in the Great Fire, Graunt was a master of assembling and making sense of ordinary information. He laid the foundation for the ways of categorizing, counting, and rendering facts and figures that would later change the way people thought about the connections between health and the surrounding world. In 1662, Graunt published *Natural and Political Observations made upon the Bills of Mortality,* a short book that summed up his years of sorting and analyzing death. He looked at questions like who died and where, when, and how the person did so.[16] In immediate recognition of this work, Charles II personally recommended that Graunt be admitted to the Royal Society that same year.

Some have speculated that the urge to monitor life's events became a preoccupation in London out of sheer necessity. Graunt himself admitted, in the first edition of *Observations,* that he did not understand why he bothered to catalog the patterns of illness and deaths "having (I know not by what accident) engaged my thoughts." But he also pointed out that these records contained valuable information that was then being discarded. "There is," he confessed, "much pleasure in deducing so many abstruse, and unexpected inferences out of these poor despised Bills of Mortality; and in building upon that ground, which hath lain waste these eight years. And there is pleasure in doing something new." He intended his work to provide the government and church with a rational way of setting policy, whether that required counting the number of fit fighting men or new, surviving babies.[17]

Graunt counted and organized people by sex, state, age, religion, trade, rank, degree, and how and where they died and what sickened them. In his system, deaths by "cancer, dog bytes, drowning, plague, fryght, childbirth, feaver, head-mould, rupture, scurvy, spotted feaver, stone, stopping of the stomach, stangury, teeth, ulcer, wormes, French pox [the British term for syphilis], small pox, and burnt in his bed" were recorded weekly.[18] He reported on christenings and burials of males and females.

Whether Graunt ever profited from his tabulations is not clear. His little book went through at least five editions and provided the basis for understanding that luck alone did not determine when people lived or died or what illnesses or other misfortunes they suffered. It was possible to calculate, on the basis of known facts about a person's life, how long and well they would live and even the likelihood that they would die of a given disease. Graunt even allowed that the "prevalence of acute and epidemical diseases might give a measure of the state, and disposition of this Climate, and Air . . . as well as its food."[19]

You could say with little exaggeration that the conditions in London, with its growing numbers of residents and early commercial activities, forced the invention of public health statistics. As early as the seventeenth century, Graunt's approach allowed businessmen to create tables of probability that laid down the odds that specific events would happen.

The impetus for counting people did not just come from the clever work of a single businessman. It stemmed from something quite practical—the need to raise money for governments and churches. When the king's or archbishop's finances depended solely on how much produce was farmed, it did not really matter how many people there were. The more crops, the more funds. But in urban London, there were few farmers and it was impractical to collect a percentage of the various goods each person produced. Moreover, some people, such as servants, produced nothing collectable at all. For these reasons, collecting taxes required the counting of heads. The information assembled by Graunt and others became the basis for one of the earliest and soundest systems of true social security. Cities in England and elsewhere began to finance basic services through the sale of life annuities. At a time when the average person lived to age thirty-five, selling people the chance that they would collect money after age forty was such a sure thing that governments bet their funding on it. The Dutch began depending on such a system at about the same time, with many towns staking their revenues on annuities sold. In London, you could buy insurance against numerous fates, from death by horse to loss of virginity. But no one ever sought a guarantee against fog.

At the same time that Claude Monet was portraying the beauty of London's unique fogs, a British veterinarian reported a mysterious incident. Most of the prize cattle that had been brought from the countryside to London for Cattle Show Week in November 1896 suddenly fell seriously ill and had to be slaughtered. In his description of this episode a century later, David Bates reports that "the post-mortem appearances were indicative of bronchitis; the mucous membrane of the smaller bronchial tubes was inflamed, and there was present the lobular congestion and emphysema which belong to that disease."[20] Basically, the cows' lungs were swollen and full of fluid.

Several writers asked whether these prized cattle had somehow been more vulnerable because they had been reared in the pristine country air. Perhaps their lungs lacked the strength of cows more accustomed to city conditions. Some weeks after the episode, a medical journal declared that the deaths defied explanation: "If any one, a few weeks ago, had suggested the possibility of a London fog doing serious damage to cattle, or other animals, submitted to its influence, he would have been looked upon as supplying in himself a melancholy instance of intellectual fogginess."[21]

Many cows of less distinctive breeding, brought from the same pristine air by the same farmers, did much better. Why? Years later, investigators deduced that these inferior beasts were protected because the straw in their stalls was not changed as often. As my playful young son used to insist, it is possible to be too clean. But we are getting ahead of that story.

In 1952 there was still much debate about whether the Donora and Liège smogs had been strange weather anomalies. That winter in London, the debates ended. On December 8, cool air from across the English Channel settled over the Thames River valley and did not move. London's eight million residents did what they had been doing for centuries: They huddled indoors and warmed themselves by their coal stoves. Smoke ran like tap water from a million chimneys. In the motionless air, the hot vapors chilled and, instead of rising, settled back to the ground. The smoke became so thick that visibility dropped to near zero. A steam ferry rammed a vessel anchored in the river. Trains and cars crashed. Road traffic crawled, with cars led by people on foot, or just ground to a halt.

Elinor Grace Jones, then a pediatric nurse in Great Ormonds Street Hospital, remembered that the fog was so thick, "we could not see across the street. It was the blackest and worst fog of any that I had ever seen. We wore masks inside the hospital. We had to change them every five minutes. They became so black. Even though I had quite a good knowledge of the streets of London and lived in the adjacent nurses' residence a few hundred meters away, I could not find my way home."

Again, the poor cows of London were hit hard. "There was the Smithfield Show," Jones recalled, "where the prize cattle were shown at Earl's Court near the center, after being fattened, dressed up, brushed and brought in from the countryside. A lot of these animals just keeled over, their tongues hanging out like dogs." The smog was destroying the animals' lungs. As they had done a half-century before, however, the ordinary cattle fared much better.

The fate of the prize cows in this smog, we now know, may have resulted from their special status. Coal smoke is full of acids containing free ions of hydrogen. These ions can make acid rain strong enough to wilt plants and flowers. Ammonia, the key smelly component in urine, counteracts acid. The prize cows had their bedding changed every day, so their urine did not get the chance to release the neutralizing ammonia. Because nobody minded if the stalls containing ordinary cattle stank a bit, these common cows, protected by their clouds of ammonia vapor, survived the Smithfield Show when their more distinguished cousins did not.

Victor Judge, who grew up in southeast London, remembers how these "lovely still November days would unfold. The day would start with a mist. The sun would slowly be obscured. By the afternoon, we got what we used to call a pea soup, where you could not see for more than about ten or fifteen feet. The whole place came to a standstill. The only way to drive was with somebody walking in front of the car with a flare. These pea-soupers would last several days. All that coal smoke was trapped below the clouds and just could not get away."

Tony Fletcher, a professor at the London School of Hygiene and Tropical Medicine, recounted how a friend's father made it home during the 1952 smog: "His dad had just about given up and had sat down on the sidewalk in dismay. He could not really see the street signs and had no idea where he was. Suddenly, he heard a sound, but could not see where it was coming from. It was a tap tap tapping of the stick of a blind man, making his way down the street. In the end, the blind fellow walked him home."

John M. Last, today one of the leading figures in public health but then a young physician working in Edinburgh, happened to be in town that day. He recalls not being able to see his own feet as he walked along Piccadilly. Subsequently, Last worked in a number of London facilities during other smogs but remembered none as bad as that of the winter of 1952.

Physicians who had heard of the Donora smog would have found London sickeningly familiar. Again, the undertakers ran out of caskets. Florist shops exhausted their supplies of funeral wreaths. Hospitals ran out of rooms and beds and filled their halls with patients on stretchers. Emergency rooms overflowed with stricken youngsters, the elderly, and otherwise healthy people whose lungs could not abide the toxic smoke. Temporary morgues were set up at several key points in the city to deal with the sudden influx of corpses. London schoolchildren from that period recall eagerly waiting for the opaque haze to become dense enough that they were unable to see across the street. Once this happened, by some perverse logic, they were discharged by officials who reasoned that it might hamper public safety and the ability to keep traffic under control if they were to stay in school the entire day.

"Mystery killer fog!" made front-page headlines. But while Donora was a small town that could be quickly forgotten, London was then the world's largest city, a center of commerce and culture. In sheer scale, this disaster could not be ignored. In one week alone 4,703 people died, compared with 1,852 during the same week the previous year.

Bates recounts the reluctance of government officials to accept that so many people had suddenly dropped dead merely from breathing dirty air. He pithily adds: "The public realized this earlier than the government of the day."

One member of Parliament put this episode into context when he asked Harold Macmillan, then the minister of housing, "Does the Minister not appreciate that last month, in Greater London alone, there were literally more people choked to death by air pollution than were killed on the roads in the whole country in 1952?"[22] This question effectively states the basic challenge to environmental health. A sensational and unexpected catastrophe that affects a group of people at one time provokes attention and sometimes a political or regulatory response. In fact, the member of Parliament did not quite get it right: Not all those who died in the week of the fog succumbed to air pollution. Some deaths are inevitable. But the number of excess deaths for the week was an estimated 2,800—a high number by any reckoning.[23]

"Members can't blame my colleagues for the weather," Macmillan replied.[24]

The centuries-old call to curb coal fires was raised yet again. But even in the face of such an enormous death toll, and private entreaties from the royal family, Macmillan resisted pressures to reduce such burning or improve efficiency of current plants, contending that this could not be afforded. "I am not satisfied that further general legislation is necessary at present. We do what we can. But the Honourable Gentleman must realize the enormous numbers of broad economic considerations that have to be taken into account."[25]

The country was operating under phenomenal economic pressures. The government had begun selling horse meat for food, because of inadequate supplies of other sources of animal protein. Rationing of food did not end completely until 1954. Candy only became available in 1953.[26] The "broad economic considerations" to which Macmillan alluded were these: Britain faced a massive war debt of more than £31 billion, as well as growing expenses for what was becoming the Cold War. To raise foreign exchange to pay off its debts, the British government was reportedly selling the cleaner-burning coal reserves to U.S. and European businesses and keeping the dirtier coal for use at home.

Like the Donora disaster, the London fog gave rise to some official studies. But as far as Macmillan was concerned, all these efforts were for show. A confidential memo, uncovered in the early 1990s but written shortly after the lethal smog of 1952, shows the housing minister's breathtaking cynicism: "Today everybody expects the government to solve every problem. It is a symptom of the welfare state. For some reason or another 'smog' has captured the imagination of the press and people. . . . Ridiculous as it appears, I suggest we form a committee. We cannot do very much, but we can be seen to be very busy, and that's half the battle nowadays."[27]

Macmillan proposed taking actions that would make it look like something constructive was underway: "There are some short term things which we can do. We can gain popularity by them. The masks etc."[28]

Some U.S. tobacco companies had offered to provide cloth masks, stamped with their brand, but this was rejected. Macmillan instead ordered the National Health Service to provide for the distribution of three million of Britain's own useless face masks, rather than provide free advertisements for U.S. firms. By around 1957, it was clear that the masks had done nothing.

Smogs hit London repeatedly during the period from 1952 to 1962, and each incident was associated with higher-than-normal death rates. During one thirty-six-hour period in 1956, one thousand more people died than would be expected under normal conditions.[29] Bad air was again involved.

Last remembers how much things had changed by the time he returned to London in 1961:

> We lived at Newington Green, and every day I walked down through Whitechapel—Jack the Ripper's patch—to the London Hospital Medical College, about a four-mile walk. So I know that part of London like the back of my hand. It had still been very grimy in the 1950s, even on fine sunny days (which are frequent in spring and summer). By 1961, the Clean Air Act had transformed London and other previously smoggy cities like Manchester, and the pervasive acrid smell of coal smoke had gone.

John Evelyn's insightful and passionate arguments, it turns out, were far ahead of the science of his day. But he could never assemble his information in a persuasive way, and his passion may well have interfered with his being accorded the credibility that science required even then. In 1987 the enterprising Peter Brimblecombe pulled Evelyn's argument together. He found dates of "Great Stinking Fogs" recorded in the daily weather diary of the astrologer John Gadbury in 1668 and 1669, and compared them with deaths recorded in the city of London during that same time period. Employing facts that have been sitting in record books for nearly three centuries, Brimblecombe showed that the highest death rates at the end of the seventeenth century in London occurred during the weeks following these notable fogs. Evelyn was right all along.

Some traditions die hard. Some do not die at all. What Brimblecombe showed for London of the seventeenth century also applied in the Great Killer Fog of 1952. The true toll of London's smog was hidden for years within official documents. It just did not make sense that a massive black smog hit for a week and killed several thousand people, and then everything quickly returned to normal.

What does "normal" mean in this situation? How can we know what is the usual rate of disease in a population? That is where the tools of epidemiology get honed and refined. We ask how many people of what ages usually go to hospital or clinic with respiratory problems. We look at records over long periods of time. We consider how many die every day from specific causes under normal circumstances. We then take these total numbers of cases and develop morbidity rates based on the ages of people in the population. With this information, we can compare rates of illness across time.

In the 1950s in London and elsewhere, public health statistics were only just starting to be retrievable for use in research. National systems for death certificates in the United States were only established after World War II. No computers made routine calculations of daily rates of death or disease. This information was instead scattered among insurance records, bills for doctors' visits, applications for employment compensation, requests for hospital beds in respiratory wards, and many other sources. Estimating how many people had bronchitis, pneumonia, or other health problems required sorting through all these sources of information.

Some anonymous researchers who first tried to make sense of the London story for the official National Health Service investigation at the time found something quite fascinating. Despite the assurances that government officials later conveyed, death rates did not return to normal for nearly three months after the smog. One early preliminary report noted that deaths and illnesses remained abnormally high until the end of March 1953.[30]

The few officials who were privy to this early assessment worried over how to handle the news. If the effects of the smog extended not for a week or two but for several months, then the problems of air pollution were worse than anyone had then imagined. An end to the burning of dirty, brown coal at home and its replacement with some form of gas or with the expensive, cleaner-burning coal then being sold for export would increase the cost of keeping things as they had been. The challenge for London was not just to tolerate the nuisance of dirty skies with the proverbial stiff upper lip (and lungs), but to assess how the health of its citizens was affected by prolonged episodes of polluted air.

It is not clear who, or what group, started the notion that all these deaths could be attributed to flu. We will never know which officials decided to apply this particular spin. But between the time preliminary data began to show persisting increases in deaths and the time the final report appeared, all these additional deaths had come to be seen as simply a bad case of flu going around London. There is only one problem with this explanation. There is no evidence to support it.

I began to think about this problem a few years ago, when I saw this pattern of continuing higher death rates. I asked several colleagues, "Do you think we could resolve this question by looking at some other records?"

Most of them nodded and allowed that the real toll of this smog had probably gone untallied. But funds are not given out for corrections of the historical record. Nobody wanted to take the time to work on that old puzzle. Finally, Michelle Bell, an enterprising young doctoral student from Johns Hopkins University, took up the challenge with me. Our procedure was, in theory at least, pretty simple. We made estimates of London's air quality in the winter of 1952–1953 and compared pollution levels with rates of death and disease (Figure 2.1).

Because the British are quite fond of their bad weather, invoking it to explain a host of national traits, we had at hand all sorts of weather information, including detailed temperature and humidity records. For information on air pollution, we relied on twelve stations that were placed throughout greater London and measured particles and sulfur dioxide gas. The standard method for measuring particles in outdoor air was basically to force air through a thin paper filter with a high-volume pump and then, each day, weigh the total amount thus collected. Measuring

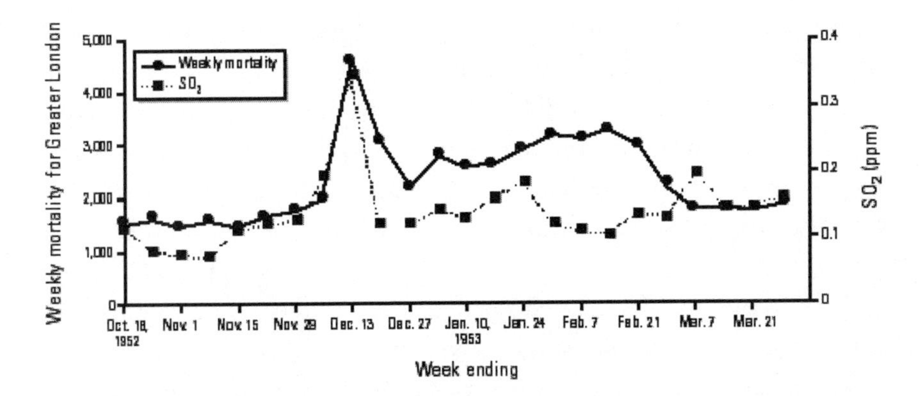

FIGURE 2.1 Weekly total mortality and average atmospheric sulfur dioxide (SO_2) concentrations for greater London. Deaths in greater London in late 1952 before the episode averaged about 1,570 per week, similar to the death rate in 1951. After the December episode, mortality remained elevated for some time. For the week ending December 13, the number of deaths was roughly 2.5 times higher than the corresponding weekly averages from 1947 to 1951. Mortality remained almost twice as high as before the episode for the next two weeks. Deaths in January and February 1953 were approximately 50 percent higher than expected, based on figures available for the same months of the previous year.

gases remained a fairly crude process, involving a number of technologies, all of which yielded slightly different estimates. Although these inconsistencies prevented us from learning the exact level of sulfur dioxide in the air on any one day, the measurements did suffice to give us a rough picture of how the levels changed from day to day.

So, how bad was it really? Of course, what was normal for post-war London still was much higher than levels found today in most modern cities. Both sulfur oxides and particulate air pollution ran three to ten times higher than even the high 1950s level during each day of the smog. The air did not return to normal for three months.

Then we estimated rates of death and illness for the entire three-month period. Although death is certain, figuring out ways to measure the rate at which deaths occur in a large city is not. We calculated death rates by taking the number of recorded deaths every day from the British General Register Office of Statistics, and dividing these by the best estimate of the total population of the town. In London in 1952, about 1,570 people were joining the angels each week. But in the week ending December 13, there were over 4,000 deaths—2½ times the normal number. This was a disaster of monumental proportions. The death rate soared that week and did not return to rates of the previous year until the beginning of March 1953. In all, about 13,000 more people died between December and March than one would have predicted from historical averages. Could this all have been due to a hidden epidemic of flu?

To find out, we also created sickness or morbidity rates, but this took a bit more resourcefulness. Because Britain had a form of national health insurance starting in 1948, weekly insurance claims for new illnesses were reported to the Ministry of Pensions and National Insurance. We had to decide what ailments, out of all those reported, could be connected with air pollution. We eliminated such things as fractures, traffic accidents, work injuries, and suicide attempts and focused on lung and heart disorders that required some form of medical treatment. To get some idea of the patterns of these diseases, we examined insurance claims, hospital admissions, and reports of pneumonia and influenza. We also included medical emergencies by taking reports from the Emergency Bed Service Bureau. Pneumonia is a serious disease for which doctors were required to keep formal records.

The British government's records of old pneumonia cases in the London area show that throughout the first two weeks of the December 1952 smog, official notices of pneumonia ran between two and three times higher than normal. Before the episode, hospitals in greater London admitted approximately 750 cases of all kinds every day—one in four of which involved lung problems. For the week ending December 9, 1100 patients were admitted—with four out of every ten having lung ailments. But for two months after the fog episode officially ended, the death rate remained at least 50 percent higher than normal.[31]

How could we evaluate the idea that the increase in ailments after the smog was due to some silent influenza epidemic? Influenza was not officially reported in any regular way, even though it was understood that flu could be deadly. We needed to find out whether flu lay behind this jump in pneumonia cases, hospital admissions, and deaths. To estimate the influenza rates, we looked at all reports we could find for clinics at the time in question. We also checked out applications for medical insurance and for emergency hospital beds.

These records gave no evidence of an unusually broad epidemic of influenza that winter. But this was not conclusive. After all, some clinicians had agreed that they had seen influenza cases and raised the possibility that this could account for much of the continued problem. How could we know that there had not been a sudden surge throughout the area?

We drew on several lines of evidence. First, the typical rate of influenza for any given winter could be deduced from historical records from 1949 to 1968. Over this period, out of every 1,000 people, about 80 developed influenza during a single year. This is equivalent to a rate of 8 percent. We also know that at this time, only 2 of every 1,000 people who developed influenza normally died of the disease—a fatality rate of 0.2 percent.

If these rates were applied to all those living in greater London in 1952 (some 8.3 million people), then more than 600,000 cases of flu would have occurred. If 2 of every 1,000 cases of those with flu died that year, then this should have produced about 1,300 influenza-related deaths for the entire year. In its final report on the London smog episode, the British Ministry of Health officially announced that 5,655 deaths from influenza had occurred in the first three months of 1953 alone. If this number were accurate, then a major epidemic would have been in full force. At a 0.2 percent death rate, almost one-third of London's population—some 2.5 million people—would have had to have gotten the flu that winter. Surely the 1952 report by the Ministry of Health's Chief Medical Officer on the State of the Public Health would have noted such an outbreak, but there is no mention of it.

Still, to be sure that flu-related deaths did not account for the continuing high rate of illness and death, we needed something more solid. And we might never have found it if not for an extraordinary coincidence. There happened to be an influenza study under way during the winter of 1952–1953, employing 12,710 volunteers from the London area. Half of them were given vaccines for the prevention of influenza. The other half served as the control group and received no vaccine. Of those who were vaccinated, about 3 percent came down with influenza—three persons out of every hundred. Among those who were not vaccinated, 4.9 percent did—nearly five persons out of every hundred. For the group in this study, the rate of influenza for 1952 to 1953 was a bit lower than the historical average of 8 percent. What this means is pretty straightforward.

If we apply the rate of influenza that prevailed in the control group (4.9 percent) to all of greater London at the time (8.3 million people), and applied the flu death rate (0.2 percent), then there should have been only about 800 influenza-related deaths in all of 1953. In fact, however, the deaths between January and March numbered 8,635 more than in the previous year. The vast majority of the excess deaths occurred in January and February. If only 800 to 1,300 deaths in London could have been due to influenza for the entire year, this means that 7,000 more people than normal lost their lives.

The official report of the London fog of December 1952 was issued in 1954. It included a graph that looks just like the one we produced, showing the same high rates of death, but with one addition. Right under the curve of continuing high rates of deaths in January and February, was inserted the phrase "INFLUENZA 1953." Would that merely inserting the word *influenza,* even in capital letters, could make it so. We have found nothing to indicate that such an epidemic really was going on and could have accounted for the patterns that plagued London for several months after the lethal episode of December 1952. In fact, Report Number 95 states that there was no evidence of flu in London at the time. It further admits, "It would be unreasonable to assume that deaths resulting from the fog ceased abruptly on or about 16th of December."[32]

Despite this admission, the report focused only on the deaths and ailments that occurred within the two weeks of the smog. A detailed appendix to the report showed the numbers of deaths every day in every county in every administrative zone, from November 30 to December 16. To support the view that increased deaths after mid-December were due to flu, the report invoked a lone study from the south of England in January 1953, showing some cases of influenza among those with pneumonia. The only way that such an epidemic could account for so many deaths would be if the rate of flu had run four times its normal rate of 8 percent. No papers from the time suggest that such a rampant epidemic was under way. The World Health Organization provided a global overview, noting that reported cases of flu for 1952–1953 appeared relatively mild, with local outbreaks caused by influenza virus B, whereas broader epidemics are typically tied with influenza virus A. If anything, reported deaths from flu in London and other cities had been higher the year before than in 1952–1953.[33]

This was not the first time that the official report of a pollution episode got it wrong, nor would it be the last time that people altered a preliminary report to produce a version more pleasing to those in power. Like the officials in Donora, the London officials placated authorities who were pressed by the economic need to see the use of domestic coal continue. The approach of studying a public health disaster in order to find ways to downplay its importance relies on a well-known method of proof that recurs throughout the history of environmental health: where the evidence does not conform with what people would like to believe, just make sure that the facts can be adapted. This adaptation can take many forms. Sometimes no studies are undertaken at all, and assurances are given that whatever has happened was just an act of G-d or a freak of nature. Sometimes the times and places being studied are altered so as to make differences harder to see. The places looked at may be expanded so that big differences between smaller regions cannot easily be seen; or, as in London, the time being studied may be chosen so as to exclude the longer-term impacts altogether. Other times, people simply stop looking. Eager to find an answer that would not disrupt Britain's war recovery efforts, the public health service and the government of the country moved on to the big challenges of putting the country back together.

For centuries, nobody paid attention to those who advised about the dangers of bad air. Nobody took John Evelyn's ideas seriously—or seriously enough. Those who were half-persuaded by his observations were apparently overruled by more immediate concerns. The pronouncements and warnings of royal commissions and stern churchmen got nowhere. Sooty buildings, dirty shirts, and the occasional death were understood to be the price of progress.

Thus the connections between dirty air and health remained hidden for years. For every John Evelyn prepared to warn of such hazards, there were any number of Harold MacMillans unwilling to be diverted and eager to dismiss their warnings. A massive flu had not really hit the town the winter after the smog of December 1952, but the idea of an epidemic proved more acceptable than other alternatives. For the British leaders of the time, it was as much of the truth as they could handle.

3

HOW TO BECOME A STATISTIC

I don't care about pollution,
I'm an air-conditioned gypsy.
That's my solution.
—The Who

In death as in life, my Uncle Len was surrounded by doctors. He collapsed while playing his usual competitive round of handball with a friend who was a physician. Two doctors set to work on him right away. When my Aunt Irene arrived at the emergency room in Cedars of Lebanon Hospital in Los Angeles, the doctor who had been my uncle's handball partner told her to wait. They were still working on him. They just needed more time.

Time, unfortunately, had run out for my robust fifty-year-old uncle with his twenty-nine-inch waist. My aunt remembers the surreal scene. "When I saw him at the hospital, I could not believe he was not alive. His skin looked normal, almost rosy. I shouted, 'He can't be dead! He looks better than he has in days! His color's back.'"

That December day in 1969, my uncle was not the only resident of Los Angeles to die. The air contained enough carbon monoxide, sulfur oxides, and particulates to keep some people indoors. Not Uncle Len, who was intent on his usual vigorous workout. During this period of air pollution in Los Angeles, the city of more than 2 million had about one hundred deaths every day. For every person who died, hundreds more would develop chronic coughs and thousands would experience what we scientists call a restricted-activity day. This is a day when you simply cannot go about your normal business. Eyes burn, throats ache, lungs cough. People cannot see straight or think clearly, because their heads are clogged. As for my uncle, he had the ultimate restricted-activity day—a heart attack. That is what his death certificate said. That is what thousands upon thousands of death certificates said for folks who happened to succumb during times of high pollution. Pollution itself never shows up on death certificates. With rare exceptions, the diseases worsened by bad air are among the most common afflictions in the modern world: heart disease, cancer, and asthma.

The real lesson of Donora and London is not just that brief, intense episodes of visible air pollution from industrial sources or coal fires can quickly fell the weak. It is that daily exposure to low levels of pollution that could not be seen or smelled can ruin the health of millions. The new science of toxicology that arose with the industrial era rested on the notion that dose made the poison. In other words, substances that could kill at high doses could be expected to be harmless at lower ones. Dispelling this assumption of the universal safety of lower doses took many years of work by dedicated, innovative, and sometimes lonely scientists.

So Uncle Len had a heart attack. Everybody said heart disease ran in the family. It sure did. Uncle Len had been a very fast walker, even a runner, wherever he lived. He even ran in Donora, when he was not jumping from rooftops or horsing around during its dirtiest old days. He ran in Los Angeles all through the boom times of the 1950s. All those years of exercise gave him a well-muscled body that fit right into the Southern California culture. Uncle Jack, Aunt Ruth, Aunt Gert, and my mom did not run at all. But they all developed some form of cardiovascular ailment. Their mother, Bubbe Pearl, who had arrived in the Monongahela Valley when she was twenty, had a heart strong enough to withstand two dozen heart attacks. But not the twenty-fifth.

The heart remains the one muscle on which all others depend. By any definition a chest spasm, or myocardial infarction, is a medical emergency. Whether this spasm comes from inhaling carbon monoxide or from a block in one of the heart's two main arteries, blood that normally sends oxygen coursing throughout the body is severely reduced or cut off, killing muscle in the process. The longer the interruption of blood, the greater the damage. In the days before he died, my uncle's blood may have been starved of oxygen. All that running, swimming, and ball-playing meant that with each of the 20,000 breaths he took every day, he was inhaling pollutants that made his body work harder to keep the blood flowing.

In the 1950s, people were moving to Southern California at the rate of more than a thousand a week. Like the hundreds of thousands of newly arrived immigrants to this promised land, Uncle Len worked some distance away from where he lived. Every day, he joined the expanding trail of suburban residents who drove into the heart of the most polluted zones of the time, spending hours in traffic before he arrived at the city center. Back then, people felt that being inside a car protected you from the bad air outside. We now know that levels of carbon monoxide and particles in cars during travel can be ten times higher than outdoors. You can be an air-conditioned gypsy, live and work in air-conditioned rooms, and still not escape invisible gases and particles.

The hemoglobin within blood cells combines with oxygen to keep the human body properly oxygenated. Gases like carbon monoxide quickly attach to hemoglobin and thus block the ability of blood to carry oxygen, creating a kind of iron deficiency or anemia. Normally iron is bound up with hemoglobin to give blood cells their rich, red hue and give the skin of Caucasians its rosy tinge. The old noble families who lived in fire-heated castles filled with invisible, odorless fumes of carbon monoxide were called bluebloods for a reason: Like my well-exercised uncle, they tended to look blue-gray. At the moment of death, blood from the heart flushes to the skin and no longer gets pumped anywhere. That is why Uncle Len looked better dead than alive.

The Los Angeles basin is an area of approximately 1,630 square miles enclosed on three sides by mountains and especially prone to stagnant air. Even though the surrounding hills are not much more than one thousand feet high, winds from the Pacific Ocean, such as they are, are usually gentle and variable. By 1955, 5 million people lived in this basin and half of them had a car.[1] Each day, they burned about 58,000 tons of natural gas, fuel oil, gasoline, and garbage, releasing more than 3,000 tons of air pollutants and blanketing the Southern California mountain ranges with up to 20 pounds of nitrogen compounds per acre. They put enough nitrogen into the air that it saturated the creeks, leaving them with levels of nitrates 50 percent higher than what was then considered safe.

The history of Southern California cannot be separated from the history of the car. In the 1940s, automobile, energy, and tire companies had a vision of the big, mostly flat land as a blank slate primed for lots of big roads. Henry Ford had offered a straightforward solution to growing urban poverty: "We shall solve the city's problems by leaving the city."[2]

The grand design for paving the United States originated with several far-sighted American business leaders who saw in the expanding cities a phenomenal economic opportunity. By the early 1920s, the U.S. automobile industry had grown stagnant. In 1921, General Motors (GM) had lost $65 million. Bradford Snell, former U.S. Senate counsel, noted that at that point, virtually every city in America with more than 2,500 people had its own electric rail system. Nine out of every ten trips by vehicle were taken by streetcar; only one in ten Americans owned a car. The country maintained 1,200 electric streetcar and intercity railways. On 44,000 miles of track, some 300,000 employees served 15 billion passenger trips, and generated a revenue of $1 billion each year. Meanwhile, the car market was saturated. In 1922, Alfred P. Sloan, Jr., created a special unit within General Motors expressly charged with supplanting U.S. streetcars and other electric rail transport. The world's largest intercity rail line was the Southern Pacific, owner of Los Angeles's famed Red Car trolleys. The Red Car line included 1,500 miles of track that extended 75 miles north from San Bernardino to San Fernando, and south to Santa Ana.[3]

Snell's report to the Senate in 1974 on the automobile industry's efforts to discourage public transit documented the monopoly practices of a consortium of firms over a period of decades, starting in the 1930s. GM, Goodyear, Standard Oil, Phillips Petroleum, Firestone Tire and

Rubber, Mack Truck and a few others had set up shell companies in most major cities. These companies went about buying up and shutting down smaller transit systems and converting them from rail to bus. By the mid-1950s, these companies had motorized more than 900 of the nation's 1,200 electric railways. The Los Angeles metropolitan area comprised two companies, Los Angeles Railway, with 1,042 yellow streetcars, and Pacific Electric, with 437 red electric trains and 1,500 miles of track. Within a month after it was acquired by American City Lines, a front firm for General Motors, the Los Angeles Railway announced plans to scrap most of the streetcar lines. By 1953, a number of them had already been acquired and destroyed. In 1954, at the former streetcar barn on the west side of Los Angeles, huge bonfires were lit as kerosene-soaked streetcars and electric trains that formerly served Hollywood were burned. A few were placed in museums, relics of an era when most folks took public transit to get to their jobs.[4]

Not all this effort went unnoticed. In 1947 the Truman administration filed two charges of conspiracy to violate the Sherman Antitrust Act against ten of the largest U.S. corporations. They were indicted by a California grand jury for conspiring to eliminate competition from transit companies and for cornering the market in motor buses. E. J. Quinby, president of the Electric Railroader's Association, charged the railroads with "an organized campaign to deprive the American public of their splendid railway systems."[5] On March 12, 1949, General Motors and the nine other corporate defendants were convicted on one charge of monopolizing the sale of motor buses and supplies to National City Lines Companies. They were fined five thousand dollars—the price of two small cars.

Undeterred by this momentary setback and by adverse publicity tied with their antitrust conviction, the car and oil firms, construction companies, and others forged ahead. California's cities got rid of streetcars and their heavy metal tracks that hindered cars from moving smoothly through the streets. Across the United States, most electric trolley lines shut down, creating a gap in transport and a ready market for millions of cars, trucks, and buses.

Southern California grew as the promised land, where everyone had a right and a need for a car. Roads were built at a dizzying pace. In

1947, nearly two out of every five workers used public transportation. Two decades later, fewer than one out of ten did. Today the number is fewer than one in twenty nationwide. From 1950 to 1970, the number of vehicles in Southern California tripled, the population doubled, and the miles of road built grew more than 50 percent.[6]

As a city girl from the Bronx, newly married at age twenty, Aunt Irene loved the brand-new Pennsylvania Turnpike. She hated the dirty, smoky air of Donora, where she and Uncle Len lived for the first few months of their marriage after he was discharged from the army. The turnpike was her link to civilization. Irene and Len regularly hitchhiked back and forth to New York, riding with big twelve-wheelers or traveling salesmen. In 1950, they piled their belongings into a car and joined the exodus from the Monongahela Valley to the romantic edges of Los Angeles. At that point, few folks realized that a city full of fast cars and big roads also emitted hundreds of thousands of tons of pollutants.

At the same time that my young aunt and uncle were heading across country, Mary Amdur was developing a visionary career of research. She invented a novel method for measuring toxic chemicals that could be neither seen nor smelled and for studying how they affected animals and humans under usual conditions. In 1950, the standard test of a toxicant was simple and direct. If the exposed animal did not drop dead on the spot, the agent being studied was considered safe. Amdur wanted to know the long-term impacts of lower levels of exposure to things that do not kill animals right away, but can have more subtle and insidious effects.

Having grown up in the Monongahela Valley when mills ran at full tilt all day and all night throughout the war years, Amdur had firsthand knowledge of pollution. As a teenager, she had seen her adored young father die a terrible death at home from lung cancer at age forty. He had worked as an engineer in one of the booming coal-fired power plants of the region. Every day that they went to work, these valley men, who regularly wore white shirts during Amdur's childhood, would take a second one to wear in the afternoon. The first one would become grayed and soiled by lunchtime just from the air. Shirts could be changed. Lungs could not.

At a time when few women attempted to gain scientific training anywhere, Amdur completed her undergraduate degree at the University of Pittsburgh and her Ph.D. in biochemistry at Cornell University, spending three years at each institution. Her son, David Amdur, notes that her teachers often suggested that Mary become a medical doctor. Having witnessed her father's painful demise, Amdur insisted, "I could never do that. I could never take care of dying people." Instead, she took a prized post at Harvard Medical School as an assistant professor, working with the country's top pulmonologist, Philip Drinker, the inventor of the iron lung.

There she turned her sights to learning how regular exposures to the sorts of toxins her father had regularly encountered affected the lungs. Early in her career, by the end of the 1940s, Amdur had already shown that she would be an innovator in the new science of toxicology—the study of poisons.

The use of poison gas during both World Wars I and II had made researching the acute and immediate impacts of toxins more than an academic exercise. Toxicology got its start looking chiefly at how much of any substance could kill people and devising antidotes to block these toxins. As originally developed in 1927 by J. W. Trevan, the first test of a compound's toxicity took a small group of animals and measured the amount that could kill half of them quickly. Called the lethal dose fifty, or LD50, this test not only indicated how dangerous various chemical warfare agents could be but was also used to establish maximum amounts of drugs, such as digitalis extracts, insulin, and diphtheria antitoxin. Even though the deficiencies and cruelty of the test were clear at the outset, it remained the gold standard well into the 1980s.[7]

Amdur set out to understand the impacts of chronic exposure to agents that were not lethal. Public health experts had warned of the dangers from long-term exposures to low levels of lead, a neurotoxic heavy metal, but industry had insisted that such exposures would not be detectable. With this in mind, she devised a method that measured unseen fine particles of lead in the exhausts of automobile engines. The choice of lead quickly placed Amdur at the forefront of public health battles, where she would remain, without honors or acknowledgment, until the end of her career.

At the same time that the automobile, gasoline, and tire companies were eliminating the trolley as competition, they also set in motion plans for cornering the market on fuel. Early cars tended to have a pinging, popcorn-popping, knocking sound when running, especially when going up hills. Faster engines were said to require fuel that burned more efficiently to avoid these noises of inefficiency. Ordinary alcohol—ethanol—easily made from agricultural crops such as corn, provided a possible solution because it could boost efficient fuel burning, but it would take up about 10 percent of the volume of a gas tank. Furthermore, ethanol could not be patented, nor could its production or price be controlled by the car and fuel companies. These firms began promoting a newly invented, specially formulated compound that bound a single atom of lead to twenty atoms of hydrogen and eight of carbon. This compound's hydrocarbon chains each held four carbon atoms, known as ethyl branches; thus the liquid was called tetraethyl lead. The public already had a dim view of lead and a vague sense of its toxicity. To soften its image, the founders of the company gave their new firm a popular woman's name, Ethyl.

The Ethyl Corporation was launched in 1923 to promote and produce lead as an additive booster for gasoline. An offshoot of three of the country's largest firms—General Motors, Standard Oil, and DuPont—Ethyl Gasoline stumbled from the start. The company's own official history admits, "if ever a company started its corporate life under bleak circumstances, it was Ethyl."[8]

In January 1923, the inventor of tetraethyl lead, Thomas Midgley, sickened from his own lab work, had to cancel several speeches before the American Chemical Society. While recovering, Midgley wrote to his boss, Charles Kettering—the man who had invented the automatic starter—extolling the considerable economic potential of an additive to gasoline that could be patented, controlled, and marketed as essential for increased power. Jamie Lincoln Kitman, writing in the *Nation* in March 2000, recounted how the idea of launching tetraethyl lead as the essential fuel additive began:

> Writing from Florida in March 1923, Midgley related a mad brainstorm whose relevance had now become fully clear to Kettering. "My dear boss," he began, "The way I feel about the Ethyl Gas situation is about as follows: It looks as though we could count on a minimum of 20 percent of the gas sold in the country if we advertise and go after the business—this at three cent gross to us from each gallon sold. I think we ought to go after it as soon as we can without being too hasty."
>
> Midgley barely scratched the surface of the wealth to come.[9]

Within a decade, Kitman noted, leaded gasoline constituted 80 percent of the gasoline market and grossed about $300 million annually for Kettering, GM, and the DuPont family.

During World War I, Yendell Henderson, of Yale University, had directed the country's research on chemical warfare. Later, he developed a prototype for a gas mask for the army and insisted on being the first to test it. Expressing strong reservations about industry efforts to influence research on lead, he refused direct funding from General Motors when it was offered. Writing to the Bureau of Mines on September 27, 1924, Henderson cautioned of an impending public health disaster: "I should be willing, and even regard it as a duty, to carry out an investigation, if adequately financed and fully protected against the fact that General Motors is now deeply committed financially in reckless disregard of the possible and even probable industrial and public health hazard."[10]

Henderson's warning proved prophetic. A month later, on October 27, 1924, the front page of the *New York World* read "Gas Madness Stalks Plant, Two Die, Three Crazed." At the Elizabeth, New Jersey, plant of Standard Oil Company, workers were suffering the acute effects of lead poisoning. William McSweeney, a former general in the Irish Republican Army and an employee at the plant, was reported to have died after three days, "clamped in a straightjacket on an iron cot in Reconstruction Hospital." Management at Standard Oil explained this outbreak of madness and death at its Elizabeth plant by saying: "These men probably went insane because they worked too hard."[11]

The affected workers were producing hundreds of gallons of the acutely toxic liquid, tetraethyl lead, to be added to gasoline. These first efforts at mass production proved catastrophic. Even a few drops of lead in this fluid form penetrated the skin, coursed through the bloodstream, and ended up damaging nerves and destroying the brain.

After word of these disastrous consequences spread, the public became understandably queasy about the prospects of adding lead to gasoline. New Jersey and Pennsylvania suspended sales of ethyl gasoline. Public health officials were lining up against the additive.[12] In response, Ethyl pulled its product from the market in 1924 and asked Surgeon General Hugh S. Cummings to set up a panel of scientists to recommend what should be done.

Throughout this time, Ethyl and its owners had been devising a bold and sexy public-relations campaign. Their ads claimed that nothing less than the fate of American progress was at stake. In direct praise of the many different uses of lead as a metal in cars, one *Life* magazine advertisement in April 1923 warned that without lead, your car might run a few blocks at most. Without the lead storage battery, cars would have to be hand cranked and lights would not work. Without lead-lined tins to carry fuel, glass bottles would have to be used. Lead was the glue of the modern car—in the lights, the gas-tank solder, the radiator, the rubber tires, the wiring insulation, the switch buttons, the paint. That there were big differences in potential human harm from putting solid lead metal into car parts, as compared with placing readily absorbed lead into a volatile liquid like gasoline, went unmentioned.

In extolling the virtues of tetraethyl lead gasoline, by contrast, the advertisers never used the word lead. One image from October 1927 included a flying car with wings. If you ride with Ethyl, the ad promised, you will get the benefits of high compression and the assurance that ethyl gasoline "has absolutely no ill effect on the motor or its parts." A 1929 ad showed Ethyl herself, a healthy-looking white woman wearing overalls, with saucy lips, flushed cheeks, and arms akimbo. A handwritten, feminine inscription across the top of the ad provided a racy come-on: "Take me with you and get a kick out of driving instead of a 'knock.'" The ad explained how high-compression cars require high-compression fuel and closed with this provocative promise: "Fill your tank with Ethyl today. You'll find a real difference in driving satisfaction."

Still, there remained the matter of testing ethyl gasoline for safety. Whatever studies were conducted on lead had been carefully controlled and kept well out of public view. The Ethyl Gasoline Corporation had funded the Pittsburgh office of the U.S. Bureau of Mines to study the risks from inhaling exhaust fumes in animals.[13] The government had agreed to "refrain from giving out the usual press and progress reports during the course of the work, as [Ethyl Corporation] feels that the newspapers are apt to give scare headlines and false impressions before we definitely know what the influence of the material will be."[14] The results of this eight-month study were released on November 1, 1924, and received a front-page headline in the *New York Times:* "No Peril to Public Seen in Ethyl Gas/Bureau of Mines Reports after Long Experiments with Motor Exhausts/More Deaths Unlikely."

Midgely himself stated unequivocally that levels "on the average street will probably be so free of lead that it will be impossible to detect it." Yes, he admitted, dozens of workers had become seriously and permanently ill from tetraethyl lead and a few had died. But those fellows had been drenched in the stuff. They had been sloppy and had received doses thousands of times higher than what anyone on the streets could be exposed to. The few studies that had been done with animals showing damaged brains, deformities, and deaths after high exposures in laboratories were dismissed as irrelevant: After all, people are not animals. Frank Howard of Standard Oil asked, "Because some animals die, and some do not die in some experiments, shall we give this thing up entirely?"[15]

Why were the hazards of lead from gasoline not better understood? For several hundred years, heated or solid forms of lead had been known to injure, maim, and kill workers. As a heavy metal, lead has the same electronic charge, or valence, as calcium. Because of this, lead chemically competes with and replaces calcium throughout the body. Calcium is one of those critical materials for life that gets to go wherever it wants, except when lead gets there first. In the bones, the brain, and the blood and throughout the nervous system, all of which depend on calcium, lead can trigger irreparable damage. Putting it into liquid form, like ethyl, only increased its ability to slip into the body.

On May 20, 1925, more than one hundred representatives of labor groups, oil companies, universities, government agencies, and news organizations filled a U.S. Treasury Department public hearing auditorium to hear presentations on adding lead to fuel. Testifying were the U.S. secretary of the interior, the assistant secretary of the treasury, the surgeon general, Charles Kettering (by then president of Ethyl Gasoline Corporation), and representatives from universities, labor, and industry.

Standard Oil's Howard called tetraethyl lead "essential in our civilization . . . a gift of God." Labor representative Grace Burnham countered: "It was no gift of God for the [17 workers] who were killed by it and the 149 who were injured."

During the hearings, Alice Hamilton, the first woman ever appointed professor at Harvard, who had firsthand experience treating brain-damaged lead workers, pleaded with the chemists to come up with something safer and less toxic to reduce engine knock. The surgeon general ended the nearly seven-hour conference by announcing that a committee of experts would be appointed to look into the matter further. The committee reported back within a month that they had seen ample evidence that leaded gasoline could be safely manufactured. The official panel concluded that "there are at present no good grounds for prohibiting the use of ethyl gasoline."

As to the safety of lead in gasoline for the public, the panel called for further studies. These were never conducted.

Many folks, like those from Ethyl, did not think it was necessary or important to measure micrograms of lead in soil or in air, believing that such tiny amounts were trivial. The expert committee had advised in 1925 that four grams of liquid lead could be put safely into every gallon of gasoline. An estimated 7 million tons of lead were burned in gasoline in the United States during the twentieth century, contributing 90 percent of the lead in the atmosphere since the 1920s.[16]

Despite the view that tiny amounts of it were trivial and represented the price of progress, Mary Amdur understood that lead, once dispersed into the air as fine particles, did not go away. As one of her first assignments with Philip Drinker, she came up with a novel hand pump that could be used in the field or, for instance, in a garage or repair shop or other places where real people worked, to measure micrograms of lead in the air.[17] Some three decades after Hamilton and Henderson had pleaded with the surgeon general not to allow lead into gasoline because the heavy metal would be finely dispersed, Amdur made it clear they had been right. Formulating gasoline with liquid lead had indeed created a massive delivery system for spreading an insidious poison into the lungs, hearts, and brains of millions. Lead from gasoline would later be found to have left residues in every major city of the world where this fuel was used.

And there the matter rested for another quarter century. Like that of her predecessors, Amdur's research on lead was shunted aside, while sales of this additive boomed in the United States and internationally. Throughout its history, the Ethyl Corporation mounted successful media and political efforts to promote its product. The firm also retained numerous well-regarded and well-paid experts who argued that keeping lead in gasoline was essential for American business to grow. Whenever its executives received contrary advice—as they did, for instance, from an internal panel convened in 1942—they ignored it. Where was the proof, they demanded, that low levels were any threat at all to human health? Without independent funding of research on these questions, the absence of evidence became proof that no such harm existed.

Meanwhile, after Donora made the news in 1948, Amdur took up a second major problem. She wanted to use the tools of toxicology to understand what had caused the sudden and unexplained deaths and illness in this small mill town. In 1952, she and Drinker published two important experimental studies that looked into what could have happened in Donora.[18] These studies directly challenged one of the central tenets of toxicology—the notion that the greater the dose, the more the damage.

For their first study, Amdur and Drinker showed that the age at which animals were exposed to a toxin could determine the toxicity of certain concentrations. They focused on both very young guinea pigs and mature ones—animals known to be quite sensitive to respiratory toxins. Unlike other rodents, which breathe only through their noses, guinea pigs can be made to breathe through their mouths—just like people—forcing pollutants more deeply into their lungs. To approximate the conditions of Donora, Amdur and Drinker invented a machine that produced regular amounts of sulfuric acid mist, one of the key agents known to have been common in the Donora fog. This acid is freely created when sulfur, oxygen, and water vapor combine during coal-burning, coke production, or steelmaking, or when oils or other sulfur-containing fuels are burned. With Amdur and Drinker's machine, acid mist could be fanned into a chamber full of guinea pigs, where humidity and air flow were tightly monitored, and released in droplets just a single micron in diameter—small enough get deep into the animals' lower lungs.

Using their invention, the two scientists exposed equal numbers of young and mature animals to the same levels of pollutants. They performed LD50 tests to find how much acid it took to kill half the animals. Amdur and Drinker showed that the young animals succumbed at a dose one-third the size needed to kill the older animals. At autopsy, the lungs of all the dead animals—young or mature—were bloodied, as were some of their noses. Whatever killed these animals worked the same way in young and old. It was just that it took a lot less to kill younger animals. This work made it clear that it was not simply the dose that made the poison: The age of the animal when exposed could be even more important.

The two scientists then tackled another big question in toxicology. What happened to animals at less-than-lethal doses? It was then believed that most inhaled toxicants worked by causing spasms of the lungs or larynx, and that without this literal cramping of the lungs, no serious damage could occur. Amdur and Drinker showed that lungs that did not convulse could still become impaired. They found that a smaller daily dose given over a longer time produced much more extensive and deep-seated lung damage than did a higher dose delivered for a shorter time. Moreover, the damage from this longer but lower dose appeared permanent. This pioneering work showed that a toxic response reflected two things: the amount of time the exposure lasted, and the concentration or dose of the agent being tested. Lower doses—even one-tenth the lethal amount—if administered over long periods could cause irreparable damage.

These important results gave Amdur and Drinker a benchmark for looking at similar effects in humans. Around the same time as their animal experiments of the early 1950s, they treated a small number of "healthy men of various ages" just like guinea pigs, for periods of up to five days using levels of acidity well below those found to kill the animals. I cannot be sure where these human guinea pigs came from. But it would not be surprising if graduate students, who then as now held positions of indentured servitude, had been tapped for the role. To calculate the speed and amount of air moving into and out of the lungs before, during, and after exposure to sulfuric acid mists, Amdur and Drinker used a piece of equipment called a pneumotachograph—literally, "lung measuring instrument." They also collected acid vapors deposited from exhaled breath onto paper filters to get some sense of how much acid remained in the body.

Not surprisingly, much of the inhaled acid remained within the body. Almost as soon as exposure began, the subjects' breathing became shallower and faster, even at the very lowest, imperceptible dose. This may have been the lungs' automatic defense. Short, fast breaths draw in enough oxygen to meet the body's needs, but they prevent acid-filled air from penetrating to the lungs' deeper regions. In effect, the body's response to acid mist is to use only part of the lungs and protect the rest. Amdur and Drinker noted in their 1952 paper that they had no idea what such exposures would do to people with lung or heart ailments, which lessen the body's reserve capacity. Nor did they discuss whether the breathing of low-level sulfuric acid mists, if sustained over long periods, could induce such damage.

As a senior professor and an eminent authority in the evolving field of industrial medicine, Drinker received, throughout the 1950s, direct funding from American Smelting and Refining Company (ASARCO), a large producer of sulfur compounds. Amdur, in working with sulfur dioxide and sulfuric acid, was involving herself in the company's principal emissions and the key toxic pollutants in mill towns throughout the Midwest. Since the company was funding Drinker's research in an era when such support was hard to come by, and since Drinker was supervising Amdur, ASARCO's managers assumed they held some sway over what she would publish. They were wrong.

The year after she and Drinker had established the living guinea pig as a model for testing inhaled toxins, Amdur extended this work. Again she established a wholly original method. Using guinea pigs she had purchased on her own and running the studies in her backyard over the long July 4 weekend of 1953, she and her husband determined how combined exposures to two key pollutants released by burning coal—sulfuric oxide and particles—affected the lungs. After exposing the animals to controlled amounts of these gases much like those they would have encountered in Donora, Amdur painstakingly examined their lungs. The lung of a guinea pig is about the size of a human thumb. But when carefully unpacked from the chest cavity, they can be dissected, stained, and measured for traces of chemicals and tracks of damage. With just a few days of exposure to acids and aerosols, the linings of the rodents' lungs tended to thicken and scar.

About 98 percent of air pollution consists of five substances: sulfur oxides, carbon monoxide, nitrogen oxides, hydrocarbons, and particulate matter. How do these pollutants get into the air in the first place? Many of them are natural, even essential for life. But the burning of fossil fuels can create these compounds in concentrations that can be sickening, even deadly.

Many people think of air pollution as something you can see coming out of a smokestack or tailpipe. But any material that gets into the air can become a pollutant, provided it remains in circulation. Some pollutants occur as particles, 50 times smaller than a human hair is round. The smaller the material, the longer it can stay aloft, the farther it can range, and the deeper it can travel into the fragile, spongy architecture of the lung. Other pollutants are gases that form when an element such as nitrogen or sulfur combines with oxygen. Still others can switch back and forth between being particles and gases, depending on temperature, humidity, and other conditions.

Particles can also settle on mists of liquid droplets, where they can create yet other agents. For instance, water droplets can mix with hydrogen sulfide to yield corrosive sulfuric acid aerosols.

In the presence of sunlight and gaseous hydrocarbons, pollutants such as nitrogen dioxide and volatile organic compounds often react to form still other pollutants, such as ozone. Ozone is a gas that contains three highly reactive oxygen atoms that can break down cell walls, thicken mucus, and gum up the delicate tree of airways and passages in the lungs. Those who have asthma, when exposed to ozone, experience a sensation much like breathing through a straw. Try it sometime, and you will see why on days when ozone is just a bit higher than usual, asthma sufferers keep their inhalers within easy reach.

Consider carbon monoxide. Any time carbon is burned, whether in wood, coal, gasoline, or garbage, it binds with oxygen. Sometimes carbon bonds with a single atom of oxygen, in which case carbon monoxide results. Other times a single atom of carbon connects with two of oxygen, producing carbon dioxide. Both forms are invisible; both cannot be smelled. At high enough levels, both forms can kill you. But it takes a lot less carbon monoxide to do the job. The first symptoms of carbon monoxide poisoning in adults resemble nothing so much as flu—lethargy, nausea, tingling, perhaps an upset stomach. Pregnant women who regularly breathe air with more than fifty parts per million of carbon monoxide, either from polluted air or from cigarette smoke, will have babies of lower-than-normal birth weight.

How is it possible that gases that cannot be seen or smelled affect not only our lungs and hearts but also the size and thus the health of babies? Amdur's work laid the groundwork for understanding that anything that can be inhaled, such as carbon monoxide and very small particulate air pollutants, can reach every organ of the body. They get there the same way any other agent does, by going through the bloodstream. Once inhaled, particles that are very, very fine can slip directly through cell walls into the blood. The effect is like adding flour to gravy: the blood of people breathing polluted air can literally become thicker.

What Amdur found in her studies in 1952 was pretty straightforward: The more acid in the air, the more damage to the lungs. The smaller the particles involved, the more deeply they penetrated and the greater their impact. She found strikingly similar effects in both animals and people—heavy scarring and thickened linings deep inside the lower lung. In December 1953, at the annual meeting of American Association for the Advancement of Science, Amdur presented these findings, showing that the combined effects of air pollutants, even when not lethal, could be quite toxic. She also argued that people exposed to levels like those in the Donora smog could suffer permanent damage. Her work was clear; so were its implications. Regular breathing of acids and particulates in the air of Donora and dozens of other mill towns throughout the country could damage the ability of the lungs to function, forcing them to work harder and faster than usual.

By the early 1950s, physiologists had a reasonably good understanding of the shape and function of many of the mechanisms involved in breathing. Humans breathe by first taking air and smaller particles through the nose or mouth into the lungs. The typical adult inhales about sixteen breaths each minute while awake and about half as many when asleep. Heavy exercise or stress can increase the rate to as high as one hundred breaths a minute. Although the total volume of the adult lung is about 5,000 cubic centimeters (one half-liter), a single breath typically exchanges only about 10 percent of this volume. Even after the most challenging exercise, when a person is panting or gasping, the lungs never empty, holding about 1,000 cubic centimeters of air. Pollutants in either gases or particles enter the body with each breath. Particles 10 microns in diameter and larger are filtered out by the nose hairs; those between about 2.5 and 10 microns generally land in the mucous lining of the nose and throat. But where particles are smaller than 2.5 microns, then the body has no special filter for keeping them out. Those who work out intensely during periods of air pollution are basically conducting stress tests on their own hearts, pulling ultrafine particles as well as other pollutants deep into the lungs. Particles this fine resemble gases in that they can slide into the bloodstream from the small sections deep in the lung.

The respiratory tree contains more than 50,000 divisions. When air is taken into the body it moves through a web of smaller and smaller tubes until it reaches, at the very end, small cul-de-sacs called alveoli. Millions of microscopic blood vessels surround each single alveolus. Here is where the rubber hits the road and all things inhaled get exchanged. Oxygen, which we need to live, and contaminants, most of which we do not need at all, both diffuse through the thin skin of the alveoli's exquisitely complicated system. The tiniest particles can also pass through the alveoli into spaces in between the cells, called interstitial spaces, and thence into the bloodstream.

If your lungs were removed from your chest and spread out in flat sheets, they would take up nearly the space of a tennis court, about 50 square meters. The internal architecture of the lungs encompasses more than 2,400 kilometers of airways—the distance from New York to Florida. Each year, the average adult breathes 7 million liters of air, give or take a few million. Whether she thinks about it or not, an active adult takes in 7 to 14 liters of air each minute, some 10 to 20 thousand liters every day. Those who work or exercise vigorously may inhale up to 50 liters per minute. If this air contains only minute levels of pollution, then large amounts of it still pass through the lungs. For those whose hearts are already damaged, whose airways are a bit narrowed by other disease, or whose lungs are still growing, regular breathing of dirty air can be especially tough to handle. Children live lower to the ground, often playing right at the level of the tailpipes of cars and buses. They also breathe faster than adults, so they can take in larger amounts of toxins relative to their smaller bodies.

Smokers' lungs are just as complicated as those of nonsmokers, but they do not have the same capacity to recover from stress or to exchange good and bad air. Lungs damaged by the regular inhalation of smoke lose their resilience. In a healthy lung, a thin layer of mucus provides a shield against damaging agents. An invisible "escalator" of cilia, or fine hairs, regularly shuttles about 100 cubic centimeters of viruses and bacteria trapped in this mucus out of the lungs every day. This escalator gets slowed and can even shut down totally when a person has bronchitis or pneumonia or is a heavy smoker. That is why sick people and smokers cough so hard and often. Their lungs are in a constant state of auto-rebellion. Because guinea pigs cannot cough, their lungs bleed more easily than those of humans.

Everybody must breathe. Even when we lose consciousness, the brain makes sure the body gets enough oxygen to keep us alive. The work of the lungs is monitored and controlled for us by the autonomic nervous system, working at a speed that is hard to imagine. Even severely brain-damaged people can continue to breathe, provided that the medulla oblongata, the breathing control center atop the spinal cord, remains intact. This respiratory control system operates like the body's own airport flight controller, constantly sensing and adjusting the traffic of breathing. If the blood accumulates too much carbon dioxide, then the center calls for faster and deeper breathing. As oxygen levels rise, the rate and volume of breathing involuntarily slow down. In detecting changes in the pace and intensity of the breathing patterns of guinea pigs and their human counterparts, Amdur was in effect recording the brain's reading of pollutants in the air.

By creating a model for studying how pollutants affected living guinea pigs, Amdur opened up a whole new world of research. Unfortunately, this was not the sort of work ASARCO thought it was funding. After she presented her findings at the American Association for the Advancement of Science meeting in 1953, word began to get out that her work was pointing to serious health hazards. Some fairly heavy-handed efforts to suppress her research ensued.

Adel F. Sarofim, a former colleague who later directed research with Amdur when she worked at MIT in the 1980s and who currently is presidential professor in the College of Engineering at the University of Utah, recalled that Amdur faced enormous pressure. At the Chicago meeting of the American Industrial Hygiene Association in April 1953, Amdur, who was a small woman, was accosted as she entered an elevator. She was on her way to a meeting room where she was scheduled to present her findings on how the lungs of guinea pigs and humans fared under polluted air. Two large, tough-looking guys wearing leather jackets got on with her. They moved in closer than was comfortable. Amdur told Sarofim that these strangers looked at her hard and called her by name: "Hey, Mary, where you going? You are not going to present that paper, are you?" Amdur did not budge. The elevator doors opened, and she stepped off, went to the meeting room, and presented her findings. She did not change a single word.

Not until years later did she speak about these efforts to stifle her work. In addition to Sarofim, she also told her son, David, and Terry Gordon, one of her last graduate students, about the incident. "Though I was not physically assaulted," she said, "I felt as though hands were about to close round my neck."

After Amdur returned from her meeting, she learned that the bullying did not stop with muscle men on an elevator. Whatever was said to Drinker we can only imagine. But clearly, someone big and powerful got to him or to Harvard. Surely, phone calls were made. Perhaps private meetings were held. A batch of telegrams and correspondence that Amdur kept from this period provides evidence of intense negotiations regarding precisely what would be published about this work.

David Amdur remembers that when he graduated from college, his mother told him just how vulnerable her work had been throughout her career: "You know, dear, you were raised on soft money." Hard money was guaranteed and came with tenure. The mostly male professors with whom she worked could rely on it. Soft money, of the sort that supported Mary, had to be earned over and over again through grants and contracts.

According to Terry Gordon and Daniel Costa, her graduate fellows at MIT, Amdur told them that when she returned to Boston, Drinker ordered her to take his name off the paper that the two of them had jointly prepared on the work she had just presented in Chicago. He also told her to withdraw the paper from the *Lancet*, where it had recently been accepted for publication. When she refused, her position with Drinker was eliminated. To this day, no one knows what happened to the paper. The *Lancet* never published it, and it seems to have disappeared.

In 1953, after nearly a decade of working with the eminent Professor Drinker and producing work widely recognized as novel and important, Amdur suddenly found herself unemployed. The scientific community can be especially tough on those who step outside the bounds of whatever is considered normal behavior at the time. In the 1950s, American industry supported research with the explicit, written understanding that sponsors would control what got published, when it was released, and whether the research the sponsors funded saw the light of day. Then as now, it was unseemly for a scientist to seem too eager for public attention. It is not OK to fuss about your work in public or to issue calls for action based on your findings.[19]

After she was fired by Drinker, she received a note from Alice Hamilton, her senior colleague at Harvard. The handwritten note, dated July 9, 1953, was found among Amdur's papers after her death in 1998. Hamilton expressed sympathy for the rough manner in which Amdur had been treated, but also conveyed understanding for the position in which Drinker had been placed: "The trouble with this branch of medical science is that it is always tied up more or less with somebody's pocketbook—Maybe the companies, maybe the insurance people, maybe the doctor in charge . . . Looked at that way, realize that Philip Drinker has wife and children who are 'hostages . . . to fortune, an impediment to all great enterprises, whether good or evil.'"[20]

John Spengler, another scientist who trained later at Harvard, recalled watching Amdur work in the 1970s:

> I used to think that in science, it was sufficient to publish an important finding once. Amdur's work showed how foolish this notion was. At every step along the way, people tried to pull the rug out from under her. In fact, she got it right years before the rest of us. The world only caught up with her several decades later, by which time so many people had confirmed what she found that it could no longer be discounted. . . . It is not enough to be right. You have to publish your findings on something subject to controversy over and over and over again.

The greatest divide in the academic world is defined by tenure—in effect, a lifetime appointment to a professional post. The tradition of tenure is meant to ensure academic freedom. Those who have it cannot be fired for what they publish (or, for that matter, for failing to publish anything at all). Those who lack it must struggle to procure appointments that are temporary or liable to become so on short notice. Tenure is granted to those who, by their work, have earned permanent inclusion within the core of the academic community. At least that is the theory.

Throughout her career, Mary Amdur remained a second-class citizen. At the time she worked with Drinker, Harvard Medical School was notorious for keeping its (mostly male) junior faculty on hold for years, waiting to find out whether they would be awarded tenure. People working on really hot issues, and women such as Amdur, seldom got there. This most elite of American universities dangled the prospect of permanent tenured posts until those who received this honor were typically well past any conceivable prime, save the prime numbers their birthdays occasionally hit. After Amdur was dismissed by Drinker, she moved to Harvard's School of Public Health, where she remained for two decades in a series of untenured positions.

After publishing the kind of major papers that for many men would have clearly warranted a permanent post, Amdur left Harvard, still without tenure, as an associate professor in 1977. She had battled with the dean over the school's refusal to give tenure to one of the guys—her close collaborator Sheldon Murphy. At that point, she was not only a woman in a predominantly man's world, she was also too old to be placed on a tenure track. From Harvard she moved down the street to Massachusetts Institute of Technology (MIT), where she also held a string of nontenured posts, and ended her career at New York University.

Science can be an intoxicating enterprise for those who are enthralled with the work itself. But the world outside science can be brutish, peopled with those who stand ready to discredit and dismiss findings that challenge the way the world runs. In retrospect, Amdur seems like the loneliest of long-distance runners. She never gave up, but neither did she actively engage in changing public policy. The one time I met her, in her small office at MIT in 1984, she told me that she had always known the world would eventually catch up with her.

By 1989, it had. When the Society of Toxicology gave her the Herbert Stokinger Award for outstanding contributions to industrial toxicology that year, Amdur provided a feisty lecture entitled "Sulfuric Acid: The Animals Tried to Tell Us." In this talk, she laid it out. The failure to heed warnings from animal studies in the past had put millions of humans at risk from air pollution for decades.

In an obituary written in 2000, two years after Amdur's death at age seventy-six, Terry Gordon and Daniel Costa acknowledged her lifelong struggles and the heavy-handed interference with which she had to contend. She was able to continue her pioneering research largely because its impact remained as marginal as her academic status.

By the end of the 1950s, Amdur's impressive body of work on animals, and hundreds of studies by others, had clearly established the hazards of air pollution for rodents. This led to another round of sophisticated attacks. If purebred guinea pigs and rats developed thickened tracheal linings and shrunken lungs, or they could be made to gasp and pant and struggle for air or develop conditions like emphysema, chronic obstructions of the lung, and even lung cancer, this was irrelevant. People are not pigs or rats. Besides, the doses used in many of these studies were much higher than those usually encountered by people. Where is the proof that pollution threatens human health?

In the city to which my aunt and uncle had fled from the soiled airs of Donora, such proof has long been evident. By 1967, motor vehicles in Los Angeles released about 20 million pounds of carbon monoxide into the air every day. Because the entire region is surrounded by low-lying hills and sometimes stagnant sea air, the area is a natural laboratory for inversions. In an inversion, cold, stable air prevents warm gases from dispersing upward and cooling. Three major forces affect how big and heavy an inversion will be: heat radiating from the ground, solar radiation, and wind. Strong winds can keep the air from cooling, while weak winds (or none) can keep things stagnant. During the daytime, the sun usually heats the earth's surface. At night, however, the ground loses heat more quickly than the air. As a result, the air closer to the ground or to bodies of water tends to cool faster and stay lower than air that is farther above. When sunshine gets through the next day, this warms up the surface air, which can then dissipate. But where clouds or pollution keep the sunlight from reaching the ground, and where there is no wind to mix things up, the surface will remain relatively cooler and the air closest to the ground is unable to disperse.

Mountains, even those as low as the 600-foot Santa Monica Hills, send off streams of cool air that flow downhill to the valleys below them. If those valleys are full of hot fumes from cars, trucks or buses, or factories, and a cold, still air mass sits atop them, then those fumes cannot break through; they stick close to the ground until the cold air mass moves on. This is how inversions become toxic. Los Angeles, London, and Donora all faced the same problem with different ingredients.

When President Dwight D. Eisenhower appointed Earl Warren, then governor of California, to be chief justice of the Supreme Court of the United States in 1953, few realized that this would forever change the way the world looked at air pollution. A few years before leaving the governor's office, Warren had signaled his awareness of the problem when he signed into law the first state Air Pollution Control Act, authorizing the creation of an Air Pollution Control District in every county of the state, on June 10, 1947.[21] Los Angeles was then the largest county in the state, had the most cars, and was already famed for its brown air.

Out of this cauldron, Arie Haagen-Smit, a chemist with the City of Los Angeles, discovered how sunlight and cars—the two items that practically defined the city—played a key role in producing the brown haze that was becoming widely known as Los Angeles smog. Through a series of laboratory investigations in the late 1940s and early 1950s, Haagen-Smit showed how nitrogen oxides and hydrocarbons from the tailpipes of cars easily converted into a brown, murky haze. These same ingredients also combined to yield a colorless, sweet-smelling gas called ozone. The Nobel laureate Joshua Lederberg found this work so intriguing, he later nominated Haagen-Smit for the Nobel Prize.

Today Haagen-Smit's work appears both remarkably discerning and somewhat short of the mark. Like many concerned with pollution at the time, he assumed that irritation was the sole problem. He even produced maps showing that the regions of Southern California with the most pollution had the most cracked rubber hoses and the most people with tearing eyes and cough.[22] He hailed Amdur's demonstration that acids and particles in the air caused what he called pretoxic effects. Haagen-Smit called her work "one of the most promising approaches to the study of pollution levels." Still, he held to the view that the scarring and thickening of the lungs that Amdur had found in guinea pigs was chiefly a nuisance. As he put it, "the effects found at lower concentration do not necessarily represent toxic symptoms but may have to be regarded in the same class as sneezing, coughing, or blinking of the eye."[23]

Earl Warren's loyal lieutenant governor, Goodwin ("Goodie") Knight, became acting governor in 1953. Knight did not need maps to know that air pollution was a big issue for his fast-growing state. His overriding goal was to keep California's booming economy moving and thus ensure his reelection in 1954. Though knowing nothing of the physics of inversions or the complex chemistry of smog, Knight and his minions understood that the famous California lifestyle could fast become unattractive if people became fearful about the state's air. The hundreds of thousands who were migrating west to sunny California's coastal zones, like my young aunt and uncle, counted on finding warmth and palm trees. But they were beginning to equate that warmth with a brown stew of eye-stinging fumes. For Knight, the issue was strictly economic: If skilled workers were scared away, then companies might think twice about relocating to California, and if the people already living there began to support some sort of regulation of pollution from industries and cars, some manufacturers might even leave.

Pollution was becoming a hot political issue. Property values were even dropping in more polluted zones. On October 24, 1954, Governor Knight pleaded with five big oil refineries at El Segundo to shut down operations to see if this would reduce a blanket of heavy smog. When the smog lasted more than two weeks, he tapped state emergency funds of $100,000 to study the public health impacts for the Los Angeles basin. Homemakers wearing gas masks and carrying brooms demanded "a clean sweep of smog," in a public demonstration outside Pasadena's City Hall November 9 that same year.

Back when Knight had become acting governor, Lester Breslow was chief of environmental health for the state of California, working to figure out how air and health patterns were related. "The governor asked the Department publicly to answer the question—'When does smog become a killer?'" Breslow recalled four decades later. "At that point, Knight was in the midst of a campaign for the permanent job. He asked our small department to tell him what had gone wrong in London and Donora. He wanted to be able to assure his state's residents that Los Angeles was different, they were not in mortal danger." By the time of this campaign, people had begun to complain bitterly about their eyes and throats smarting. It was even rumored that Knight or perhaps his wife or children had serious breathing problems.

Now dean emeritus at UCLA's School of Public Health, Breslow at that time was a young physician just starting to create what would become the field of chronic disease epidemiology. He had never heard of Mary Amdur or her guinea pigs. But he knew plenty of people who felt uncomfortable living in the growing Southern California basin and he was well aware that Haagen-Smit's laboratory studies had produced smog under conditions eerily like those that were occurring with increasing frequency on the region's freeways.

"We knew things were bad. We all could feel it ourselves. Some days your eyes burned, you coughed, and the air was brown and thick and hazy. You could not even see across the valley, as large swaths of brown clouds just sat over the lowest lying areas."

Breslow delivered his report to Governor Knight in March 1955, shortly after his reelection. It began by reviewing the British government's official analysis of the London smog of 1952–1953. Massive deaths there had been due to unusual conditions, an extraordinarily long, stagnant inversion combined with a massive release of coal smoke. Los Angeles, by contrast, faced chronic problems with lower levels of pollutants. No coal stoves were involved, but the growing number of cars was a source of apprehension.

Knight and his staff remained convinced something had to be done. People were not able to move around as fast as they needed to, and health complaints were rising. Reaching for something that the state could do to address these concerns, the governor did what officials often do in such circumstances. He had the state create funding for research to sort all this out. Research is always needed. It is always of value, especially to scientists.

In fact, state-supported research on the uniquely damaging effects of smog on living tissue already existed. In 1954, a botanist working for the Los Angeles County Air Pollution Control District, Ruth Ann Bobrov, published an article in *Science* comparing the cells of the commonly found weed *Poa annua*, grown in a smog-free greenhouse, with those subject to regular smog. The gross symptoms of chronic smog exposure in leaf tissue can vary. Spinach turns silver. Romaine lettuce goes brown. Tomatoes become brown-black and mottled. For the poa grass, the effects depended on the age of the grass. Younger grass shows damage only at the tip, while older grass, with longer exposure, develops darkened and damaged shafts. Bobrov thought poa grass might be a good indicator of what smog could do to other living matter as well.[24]

In those days, nobody considered the possibility that harm to plants could be relevant to humans. Even the notion that lung damage in experimental animals contained a warning for human health was not widely shared. Crop damage was a big deal all on its own—in a state where agriculture was a mainstay of the economy, a very big deal indeed. Researchers developed some of the first instruments for measuring air pollution to study crop damage on farms—specifically, on one of the most valuable crops in the world: tobacco. In the Washington, D.C., area in 1939, the first measures of ozone's destructive ability were made in a tobacco field. Farmers, alarmed about "weather fleck injury" to their crops, had permitted ozone monitors to be set up about 3½ feet above the ground, about one yard below the tops of the tobacco plants. Sure enough, the higher the ozone, the more browned and damaged the plants. At the time, no one asked what this could mean for people , nor did anyone note that the monitors were located at precisely the breathing height of young children.[25]

To answer Governor Knight's question, "When does smog become a killer?" the California Health Department experts needed to understand things far outside the normal concerns of the medical community. They needed to know how winds behaved and what made weather; they needed to understand the chemistry of the internal-combustion engine; they needed to assimilate the work of Bobrov and Haagen-Smit and Amdur.

As the California Health Department and the Air Pollution Control Districts were expanding their research efforts, the federal government weighed in with a new law. When the government does not know what to do about a problem, and sometimes even when it does, passing a law is a way to create the sense that something is happening. Where the law merely seeks more information, lawmakers can rest easy that at least they have not changed things for the worse. In that sense, the Federal Air Pollution Control Act of 1955 fell into a respected tradition. It authorized the secretary of health, education, and welfare to work toward a better understanding of the causes and effects of air pollution. More research was needed, and the Department of Health, Education, and Welfare, an agency with no legal authority to do anything about the sources of air pollution, was the logical place to see that work begun.

Breslow and his colleagues faced several hurdles. They weren't sure what exactly they should measure or how to go about doing this reliably. Once they answered those questions, they would still need to estimate these factors' impact on human health. Even something as unmistakable as death contains ambiguities: Death may be simple, but the whens, wheres, and hows of calculating the death rate are not. You need to know the size of the population affected, their average age, and the relative proportion of people of different ages.

When the California Health Department started collecting information on air pollution at the state level, for situations in which large numbers of people got sick and died, it was easy to argue that there was an urgent problem. The harder task was to convince public officials and private companies that chronic, regular exposures to lower levels of pollutants could have broad and negative impacts on health.

In 1958, Pat Brown was elected governor, having campaigned on a platform promising concrete actions to fight air pollution, not just to study it. Brown presided over a state of 16 million people, with nearly 8 million registered vehicles, 2 of which belonged to my aunt and uncle.

Brown's 1959 inaugural address took the issue on squarely: "Air pollution is a statewide menace. It threatens the health of people not only in Los Angeles, but in every heavily populated region of the State. We must recognize that our attack on smog cannot stop at county lines. . . . I want to serve notice on every industry and every person involved that my administration will take effective action to protect the people of the State. People are more important than dollars."[26]

Many years later, Breslow still remembered his shock at what happened next:

> This was just amazing! We did not even get a letter. The new governor wanted to make something happen to clean up the air then and there! This time there was no phone call, except to announce that he was sending a high-powered team from his personal staff to meet with us and tell us what we needed to do. We knew we would have to do something about all these cars.
>
> Bill Coblentz from San Francisco and Warren Christopher, then from L.A., were two bright, young lawyers working for Pat Brown. [Christopher later became secretary of state in the first Clinton administration.] They came to our offices in Oakland, California, with a message directly from the governor. Out of the blue the governor demanded that we

> come up with a plan to set standards for air pollutants and to control exhausts from cars. I think they said, "We will give you six months to work this out." John Maga, my engineering colleague, and I looked at each other. Then we looked at them, and we said, "Look, this is impossible."
>
> Christopher would have none of that. He just looked at us and said in this low, but firm voice that he could not take no for an answer. It is not acceptable. We should just figure out how to make it happen.
>
> We really had no idea how to do this. But the politicians would not accept any answer except, "Here is a way that might work." I have to tell you, by the time we were finished, we had come to appreciate that without their pressing us, nothing much would have happened.

Breslow and his colleagues had a seemingly impossible task. Nobody had ever asked the sort of question Brown had put to them. He demanded not just research but an actual plan to reduce air pollution. What were the main sources of the problem? What specific things needed to be changed in order to make the air clearer?

"We were basically clueless as to the solution," Breslow said. "After all, we were researchers. But it became clear that we had to come up with something. So we went to work."

The whole effort was pretty much invented on the spot. They had a limited reservoir of information on which to draw. Nobody had bothered to gather statewide information on levels of pollutants, nor had anyone tried to tie these levels to various sources and then link them with health conditions. Science at that point was devoted to parsing the universe at ever finer scales of resolution—to finding causal mechanisms in cells, molecules, and even the constituent parts of subatomic particles. Now the political world had asked these scientists to do the opposite: to find broad connections between widely disparate fields—and moreover to come up with something that would actually make a difference in how bad the air was every day. This was uncharted territory and filled with risk. If the scientists got it wrong and recommended controlling things that were of no importance, then the entire economy could be imperiled and the credibility of a nascent field could be crippled.

Forced to put forward a plan, Breslow, Maga, and their colleagues devised a framework for dealing with air pollution that has become a standard all over the world. First, they asked, where did most pollution come from? What were the key sources of pollutants in air? Without doubt, cars stood out, but other sources were also important. Anywhere people used fossil fuels, they produced an array of mists, gases, clouds, fogs, hazes, vapors, dusts, smoke, soot, and fumes that could include carbon monoxide, particulates, nitrogen oxides, sulfur oxides, hydrocarbons, and lead.

The California air pioneers worked out with their engineering colleagues just how you could go about measuring these things in a way that could be repeated regularly. They also figured out how the recipes for making car engines and their fuels could be changed to reduce their exhaust fumes. And they came up with ways of measuring whether levels of certain key pollutants found on a daily basis were not healthful. Maga, a sanitary engineer, developed a way to reduce the total amount of pollutants that came out of a car's engine by changing the burn rate and the constituents of fuels. Teams of chemical engineers created systems for measuring smog and the things that went into it. Los Angeles had become a real-world testing ground for controlling air pollution.

One of Breslow's colleagues was John Goldsmith, a young physician who served as the state's leader in environmental health research for more than three decades and later became a distinguished epidemiologist in Israel. Goldsmith took a particular interest in the levels of carbon monoxide in the air of California's cities and in the amounts of carboxyhemoglobin (hemoglobin bound up with carbon monoxide, rather than oxygen) circulating in the blood of these cities' residents. Carbon monoxide looked like an equal-opportunity pollutant, causing all sorts of biological havoc. Drivers responsible for accidents had much higher amounts of carbon monoxide bound up in their blood. This was consistent with the view that this gas could dull reaction time and affect brain function. Rates of heart attack were higher on days just after those when carbon monoxide had been above 20 parts per million. Smokers obviously would have higher levels of carboxyhemoglobin from inhaling carbon monoxide, but Goldsmith found that people who did not smoke could have the same high levels as smokers if they lived in the more polluted zones—the same zones depicted on Haagen-Smit's map. This meant that eye irritation could be a kind of omen—far worse than a nuisance. Whatever was making people's eyes tear was exacting a far greater toll on their bloodstreams, hearts, and lungs.

Goldsmith came up with several lines of attack to show how acute events like heart attack could be tied with air quality. He began charting patterns in hospital admissions and comparing them with changes in daily measures of a number of air pollutants. Most strikingly, he looked back at the death rate in August 1955 and showed that 1,200 more deaths than usual had hit Los Angeles over a period of ten days in August, just when a nasty inversion of smoggy air had settled over the valley. The combination of air pollutants and heat was the chief suspect.[27]

Through John Goldsmith, my uncle made it into *Science* magazine, about two years after his last handball game. His death became part of what scientists call *data*—the plural of *datum*. *Datum* is a solitary, neutral term, neither male nor female. So my favorite he-man, can-do uncle, who in life was fully male, became a neutered point of statistical information. Other data consisted of other uncles and aunts and mothers and fathers, and even little brothers and sisters, all of whom died earlier than they would have during periods when air pollution was a bit higher than normal. In truth, the names of the thousands who have died from environmental contamination are rarely recorded. Goldsmith and Alfred Hexter's report "Carbon Monoxide: Association of Community Air Pollution with Mortality" appeared on April 16, 1971. This is where they showed that deaths from heart attacks and all other nonaccidental causes increased most on the days just after carbon monoxide levels were highest. Within their analysis, buried beyond recognition, is my Uncle Len.[28]

Goldsmith had found a way to prove what many suspected. He learned to look back and find death-dealing pollution episodes by pulling together individual tragedies. Although any given heart attack has different causes, the aggregate number of attacks follows big leaps and drops that are linked with changes in pollution. But a number of questions remained. Did these episodes merely cause those who were already sick to die a bit earlier? Or was something else going on? Was any level of air pollution safe?

Proof, to a scientist, does not come from politicians, however persuasive or determined they may be. Within the scientific community those who succeed tend to be reserved and cautious and stay within carefully prescribed activities. The bold, daring lone ranger may be romanticized by other parts of American culture, but that glorification does not extend to the internal workings of science. Funds are given to those who stay well within the confines of their field, not to those who push the envelope.

In the 1950s, folks like Goldsmith, Breslow, and Amdur pushed science beyond its safe boundaries. They were not alone. Several other researchers understood that air-pollution-related illnesses did not magically end with the great smogs. Despite the more than 3 million cotton masks handed out to Londoners through Harold Macmillan's scheme, supplemented by glamorous masks designed by some of London's top milliners after the massive killer smog, one thousand deaths more than normal occurred in just 18 hours in 1956. While those useless masks now look like proof of the arrogance and ignorance of some politicians, political posturing is not even the biggest hurdle scientists face. Other challenges come from the science itself. It was and is bloody hard to sort through the issues. The evaluation of air pollution was complicated by the fact that in most modern countries well into the late 1970s, many people, especially men, smoked cigarettes. Today there is no question that cigarette smoking damages the lungs and heart. Not until more than fifty different studies conducted in twenty countries all showed the hazards of smoking did the scientific debate end, with the publication of the U.K. Royal College of Physicians report on the subject in 1962, and the U.S. Surgeon General's *Report on Smoking* in 1964. For years before and after these landmark findings, the tobacco industry funded a successful, well-orchestrated, and perfectly legal campaign to undermine any studies showing how hazardous tobacco really was. This effort highlighted technical uncertainties about the methods used to assess tobacco's dangers and also impeded the ability to separate out the harmful effects of pollution.

There never has been a surgeon general's report on air pollution.

California's political process left a legacy of expectations. Science is inherently uncertain. More research is always needed. But the competing demands of public health and economic growth required that decisions be made about where and what to build roads and homes and how to move people around, despite the existence of uncertainties. Driven by the political demands of the era, California became the first state on many fronts. It was first to have an air pollution control agency; the first to set up some form of program to test car engines on a regular basis; the first to impose automotive emissions standards, which required "blow-by" valves to recycle crankcase emissions; and the first to set up a process for setting and changing the standards for key air pollutants.

With all this to its credit, the state also became the proving ground for industry tactics to delay and debunk regulatory efforts, and undermine the scientists on whose work those efforts would rely. This is a role for which California seems fated even today.

4

HOW THE GAME IS PLAYED

> Our industry has been backed to the cliff edge of desperation.
> —Lee Iacocca

Within a decade of London's 1952 disaster, the science of air pollution was clear enough. Lots of bad air in a short period could harm and kill people. Well before the passage of major laws in the United Kingdom and the United States, the big, undeniably dirty sources of pollution, like coal-fired plants, began to cut back, at least on visible emissions from their smokestacks. A number of experts, however, understood that the more insidious problems were not the headline-grabbing disasters like those of Donora and London but smaller exposures to poisons over many years.

If some scientific specialists knew all this decades ago, why has it taken so long for governments to act to curtail the slow, steady rain of pollutants on our communities? This chapter tells how the connections between lower levels of pollutants and chronic health problems finally became the grounds for a significant change in one American industry. What's scandalous is how many years had to pass; how often the results had to be replicated; how stubbornly, consistently, and brazenly some in industry fought against their acceptance; how easy it was to buy experts who would weigh in on the side of delay and block important studies from proceeding; and how many people had to get sick and die before the necessary actions were taken. The introduction of emission standards for cars, trucks, and buses came years after its time.

Many of those who would become leaders of the environmental cause cut their teeth in California. One of the most stalwart, if inadvertent, supporters of environmental activism was the gifted political strategist, California's former senator, President Richard M. Nixon. In a stunning victory for environmental rhetoric, on New Year's Day 1970 he declared the coming decade a time when "America pays its debt to the past by reclaiming the purity of its air, its waters and our living environment. It is literally now or never."[1]

In truth, the president did not think much about the environment. He kept the White House air-conditioning turned up in the summer so that he could have a fire going year round in the Oval Office fireplace (a habit he maintained even through the 1973 energy crisis). He liked the homey feeling. But stories of blinding, stinging, windless smogs were beginning to spread from Los Angeles to other areas of the country, and Nixon could take the measure of a political tidal shift before most others even sensed the current had changed.

Later in 1970, Gaylord Nelson, then a senator from Wisconsin and a leader in the antiwar movement, proposed the first nationwide environmental protest "to shake up the political establishment and force the issue onto the national agenda." "It was a gamble," he recalled, "but it worked." Denis Hayes, a young activist working for Nelson, spearheaded a global happening, called Earth Day, on April 22, 1970. The party was half rock concert—featuring James Taylor; Peter, Paul and Mary; Bob Dylan; and thousands of folk rock wannabes—and half political rally, with many local and national politicians staking their claims on behalf of what was fondly referred to as our lonely planet. Twenty million mostly young Americans, joined by millions of others throughout Europe, massed in parks, dumps, city plazas, and streets, cleaning up abandoned lots and riverbeds and waving banners on behalf of the earth.

Nixon anticipated that in his next national campaign, in 1972, he would face Senator Edmund Muskie, who had staked much of his reputation on environmental issues. In a series of well-publicized Senate hearings begun in the late 1960s, Muskie had called for action on air and water pollution, echoing the views of people like Ralph Nader, then a brash, young leader of even younger activists. At the end of 1970, with another stroke of the pen, the president played his trump card. After Congress refused in December to create an environmental agency, Nixon signed a reorganization plan creating the Environmental Protection Agency (EPA). The EPA was an amalgam of pieces of fifteen agencies that started life with a budget of $1.4 billion, 5,600 enthusiastic employees, and several broad mandates to protect the nation's air, land, and water.[2] Among its other charges, the agency had to have a rational basis for recommending specific emissions controls for engines and for setting standards for air and water pollutants. Whether the agency would be responsible for conducting independent research was never made clear.

As sustained episodes of stifling, bad air did not stay within city or state boundaries, the debate over air pollution began to shift. The same administration that had sought in other contexts to return power to the states became the champion of federal actions to control the major and obvious sources of this pollution—power plants and automotive vehicles. The Clean Air Act of 1970, passed the same year EPA was formed, included something that had never before been put in place in any country—a series of nationally established standards, called National Ambient Air Quality Standards (NAAQS), to set concentrations of pollutants that would "protect the public health" with an "adequate margin of safety" and "protect the public welfare." Initially, NAAQS were to be established for the same pollutants that the state of California had begun monitoring a decade earlier: particulate matter, sulfur dioxide, nitrogen dioxide, carbon monoxide, and photochemical oxidants.[3]

The law gave EPA 180 days to announce standards for these pollutants. The states would then figure out which of their power plants needed to reduce pollutants by what amounts and determine the mix of pollutants from cars, energy, and industry that would be permitted. Each state had to establish Air Quality Control Regions, within which it could set standards specific to local needs. The automobile and fuel companies had to figure out what needed to be changed to meet these standards.

For agents deemed more toxic, the 1970 law created regulations called National Emissions Standards for Hazardous Air Pollutants (NESHAPs). For these pollutants, EPA was required to set standards based on some rational calculation of their health risks that would "protect the public health" with an "ample margin of safety." How this margin of safety was to be determined, who would develop it, and with what assumptions was all left to EPA. Plenty of people were ready to help in this effort, some of whom were handsomely paid to do so.

Within four months of its founding, by December 1970, the agency had to tell the states what to measure; how to go about setting up monitoring systems; what sorts of programs needed to be established to reduce air pollution from coal burning and from cars, trucks, and buses; what sorts of engines would be permitted on cars entering the country and those made here; and even what kinds of fuels would be allowed. Under Governor Pat Brown, California had begun programs that would eventually require all cars sold in the state to have engines that met certain standards, along with a separate inspection and maintenance program. In addition, the state would go on to create bus and carpool lanes on major streets and highways, restrict the building of parking lots, set up a mass-transit incentive plan by employers, put controls on gas-handling operations, meter ramps on some highway entrances with bypass lanes for buses and car pools, and retrofit air pollution control devices on autos. If these measures failed to reduce pollution, the state even had plans for the rationing of gasoline.

The oil and automobile firms got it. They saw at once that if they did not act quickly, the entire nation would soon be headed in the same direction as California. Having successfully fought off regulations for years, they confidently prepared to take on the new agency. At least now they would have a single target, not fifty states to deal with.

The new chief of EPA, William D. Ruckelshaus, came to the job with a distinguished record in the U.S. Justice Department. As the first of many legally required deadlines approached for EPA to propose standards under the NAAQS, Ruckelshaus was handed a mountain of paper. His staff had plumbed California's records, consulted with scientists in other parts of the government, and compiled reports generated by the industries directly to justify radical changes in how cars and fuels were designed, sold, and maintained. Ruckelshaus remembered his astonishment at the amount of paperwork set before him and the changes they would require: "When I started at EPA, I was told that everybody had already agreed to these standards. They had been cleared through Congress, and HEW, which then ran the Air Pollution Control Agency, had developed a six foot high set of criteria documents. I was handed them three or four days before the deadline and told they were all signed, everybody agreed to them, there was no controversy; just announce them and the agency would start to enforce the new Clean Air Act."[4]

The only people unaware of this apparent consensus were the auto, oil, and coal industries. When they learned what the new standards for particles and sulfates would mean for cars, they cried foul, and trotted out their own experts, who claimed that the proposed standards could not technically be met. The Clean Air Act made it illegal to take a new car that had not been certified by EPA as meeting emission standards across state lines. The job of wrestling with the companies on behalf of EPA fell to a self-described professional bureaucrat, Eric Stork. He remembered that most of the car firms went into a state of culture shock at the notion that the federal government would tell them how to build their cars:

> My job for EPA was to assure that cars would not produce more than allowed amounts of key pollutants. Auto company executives were typically the sons and grandsons of auto executives. Their ethos was big, powerful cars, with fancy fins and chrome and all that stuff. This industry had been almost totally unregulated. Suddenly they have to face a new law and short, commanding bureaucrats like me with an unfashionable haircut, who say, "We don't care if your cars are big or powerful, or fancy or fast. We care only that they be clean."

The industry sought relief in the form of a one-year extension of their deadline for meeting the new standards for emissions. On the eve of one of the public hearings that EPA held on this matter, the *Washington Post* wrote this about Stork:

> The 48-year-old Stork [who] worked as a regulator for the Food and Drug Administration and the Federal Aviation Administration before

> joining EPA in 1970 spells out his views pungently: "The EPA is never going to build cars," he said. "It's the auto industry that has to build cars. But it's the job of the EPA to make sure the auto industry does in fact build the cleanest cars possible."
>
> Stork went on, "In this world, if you want to move a man or an organization, there is no use reaching for his throat because in our society you are not allowed to shut off his wind. In our society, if you want to move a man or an organization, you have to reach for another part of the anatomy. . . . The Clean Air Act has given the EPA the best grip on the short hairs of any industry that any administrative agency has ever had."[5]

Right off the bat, EPA operated in a high-magnification fish bowl. Its every move was followed by hungry cats of various stripes and power. On the one side, eager environmental groups sensed they were riding a rising tide of public support for action. On the other, a number of industries began to fear that their world was about to change in ways that could prove problematic. Those drafting the laws were more concerned with getting reelected than with what science could or could not clearly establish or with how the rules would work in practice. The atmosphere at the agency, Ruckelshaus once remarked, was like running the hundred-yard dash and having your appendix taken out at the same time.

EPA survived its first bureaucratic battles in large part because the public understood the need for federal actions. Ruckelshaus remembers the beginning of the new agency as the days when there were good guys and bad guys, and it was easy to tell the difference. The EPA administrator could count on fairly broad enthusiasm for his efforts and minimal interference from the White House. The agency was seen as taking charge of a problem that desperately needed addressing. Reviled when it stepped into private lives, the federal government was widely viewed as the only force that could prevent further degradation of the environment.

Still, some firms were shocked at the changed world they encountered. "Some American industrialists," Ruckelshaus recalled, "believed environmentalism was a fad, a lot of nonsense that would go away if they just hunkered down, fought, and publicly confronted us. They couldn't have been more wrong. They really misjudged the power of the environmental movement and its ability to galvanize public support. When [any company] decided to confront the agency or me, it was simple to take them on. We couldn't have invented any better antagonist for the purpose of showing that this was serious business, that the agency was serious about its mission."[6]

In retrospect, 1971 looks like the high-water mark of government sympathy to environmental causes. Politicians of all stripes accepted the need for action by the federal government. The Nixon White House formed the Quality of Life Review Committee, to keep a check on EPA. At first this task fell to the White House's own team of bookkeepers at the Office of Management and Budget. When it became clear that this looked too much as if the White House were trying to undermine EPA, the official duties for assisting the agency were handed off to a group founded around the same time as EPA, the President's Council on Environmental Quality. Nixon himself, while utterly lacking any personal commitment to the environment, had acted decisively because he knew he had no choice.

Right after President Nixon was reelected in 1972, Ruckelshaus had one of his few personal encounters with the man who had created EPA by decree. He was well aware that he would be reappointed and had expected to be asked to discuss priorities for the coming administration. He never got the chance. After Ruckelshaus spoke for a few minutes about what he hoped to accomplish in the second term, Nixon delivered his own message: "You better watch out for those crazy enviros, Bill!" the president warned. "They're a bunch of commie pinko queers!"

The meeting ended a few minutes later, with a promise that the White House would be in touch about policy specifics. Ruckelshaus never did get a chance to find out what environmental policy, if any, the president wanted. Political trouble came from another direction. The Watergate scandal began unfolding before Nixon's second term even began, and as one of the most admired politicians in the administration, Ruckelshaus was tapped in 1973 to become acting director of the Federal Bureau of Investigation and later deputy attorney general. He joined Archibald Cox in the investigation that would ultimately send Richard Nixon back to California.

With Ruckelshaus's departure, Bob Fri, who had served as EPA's deputy, became the acting administrator at the age of thirty-seven. On Fri's watch, public health's old nemesis, leaded gasoline, resurfaced for the long beginning of its last hurrah.

Some companies gave life at EPA a certain stark moral simplicity, and Ethyl Corporation was one of them. As a firm that had always managed to keep its ledgers in the black and its affairs free of serious government meddling, Ethyl fully expected to continue receiving favored treatment. In 1924, the company had procured the advice of some eminent scientists to get approval to put a known neurotoxin to the brain (namely, lead) into gasoline. Despite vigorous disagreement from public health specialists at Harvard and Yale, Ethyl's experts, one of whom was also working as a U.S. government adviser at the time, had convinced the U.S. surgeon general that there was no proof that small amounts of lead placed into gasoline would have any human health consequences. With this track record, Ethyl and many other companies regarded the new EPA as a nuisance requiring only a sophisticated public relations effort. There was only one problem. The Clean Air Act of 1970 explicitly required the agency to come up with standards on the lead content of gasoline for the express purpose of protecting car engines.

In 1973, just before Ruckelshaus left, the agency proposed phasing lead out of fuels entirely, a proposal based on the view that this toxin was a potential hazard to the brains of young children. Immediately the Ethyl Corporation went to work. It charged that this regulation lacked scientific rigor: The agency had failed to make the case that lead had harmed children's brains. The company claimed that the ordinary use of lead in gasoline was safe and released only minute amounts of inorganic and inactive forms of lead. This was a rehash of a position staked by Ethyl's original expert on leaded gasoline, Thomas Midgley, who asserted at a presentation to the American Chemical Society in New York on November 10, 1922, "If properly manipulated, [tetraethyl lead] can produce lead poisoning. But as antiknock, it is required in the proportion of only 3 cubic centimeters to the gallon of gasoline; and such a mixture can be handled in any reasonable way without toxic effects. When this lead-containing gasoline is burned in a motor-cylinder, the lead is converted into a solid inorganic form."[7]

The matter—like nearly everything EPA has ever done—ended up before a three-judge review panel of the U.S. Court of Appeals for the District of Columbia Circuit. The issue before the court was whether, despite the seemingly clear intent of Congress in writing the Clean Air Act, EPA had marshaled sufficient evidence to regulate lead in gasoline as a health hazard based on what was known at the time. Could the agency really show that lead "will endanger" public health and "present a significant risk of harm"? Could it determine whether the levels it proposed to permit (even if zero) provided the public health with "an adequate margin of safety"? All these requirements were expressed in the future tense. In effect, Ethyl argued, the law required of EPA a certainty about the future that the agency couldn't credibly claim. The panel agreed, ruling that the agency had not proved its case that it was necessary to remove lead from gasoline to prevent children from being exposed.

One judge did not go along. Judge J. Skelly Wright issued a stunning formal dissent. He argued that in light of the potential irreversible damage to the brains of millions of children, the EPA must act even when science remains incomplete and uncertain. In a formal review of the panel's decision before all judges of the D.C. Circuit, the full court sided with the lone dissenter and reversed the decision in 1976: "Regulators such as the Administrator must be accorded flexibility, a flexibility that recognizes the special judicial interest in favor of protection of the health and welfare of people, even in areas where certainty does not exist."[8]

But with this battle, the giddy days at EPA, when everything seemed possible, were ended. If the young people working at the agency thought they were going to make things happen quickly, they were in for a surprise. Every action would take much longer than the laws required, would end up costing more time and money, and would run a gauntlet of delays. The effort to get lead out of gasoline was not fully resolved until over a decade later. The last car designed to run on leaded gas in the United States came off the assembly line in 1976, but you could still buy the stuff at the pump until 1995. As this new century began, nearly a hundred countries—including most of those in the Middle

East and Africa—still used leaded gasoline, despite the universal acceptance of its dangers for children and fuel-industry workers.

One day in 1973, Fri learned from a group of EPA engineers that the thousands of emission tests the companies had provided the agency contained a gold mine of information. One of their leaders told Fri with great glee, "We've figured out a way to tell people how many miles to the gallon they will get if they drive any car. It's fast and it's cheap to do."

Under the Clean Air Act, Congress required that beginning in 1975 EPA should set emissions standards of hydrocarbons and carbon monoxide 90 percent below those allowed in 1970. By 1976, it was to set standards for nitrogen dioxide 90 percent lower than those occurring in 1971. EPA had to put numerical values in terms of grams per mile of what could be emitted. Some expert engineers working for the automobile companies assured the agency that cars simply could not be built that produced even 75 percent less pollutants. It was while looking at the extensive data on emissions from cars that EPA's engineers came up with a pretty simple way to calculate gas mileage as well.

Fri soon learned that the car industry was not at all impressed. At that point, the so-called Big Three—General Motors, Ford, and Chrysler—dominated U.S. auto sales. Each company privately and firmly argued that it could never make such information freely available. The companies had no interest whatsoever in telling the public or their competitors how far a given model could travel, or even the average for their entire fleet. This was the stuff of trade secrets. If that information became public, market edges would be endangered; stocks would tumble. The very future of U.S. business was at risk.

Bill Pedersen was then one of the brightest and most enthusiastic of all the young attorneys employed at the agency. He remembers Fri's encounters with the car companies with admiration: "The car companies were absolutely right about the law. In fact there was not one shred of legal authority for EPA to require that the car manufacturers make information on fuel efficiency public. But Fri was more right about what the public wanted at the time."

Fri knew very well that he was dealing with some of the most powerful companies in the country. These were the same companies that, with the oil and tire firms, had been convicted of conducting a vast conspiracy to dismantle public transportation across the United States, monopolizing the sales of buses to replace trolleys and ensuring that Americans remained wedded to their cars for decades to come. But Fri knew something else as well: Public opinion was on his side. The car industry had recently engaged in a number of heavy-handed crusades. When Ralph Nader had publicly lambasted General Motors for trying to cover up the dismal crash history of its Corvair in his 1965 book, *Unsafe at Any Speed,* the industry did not help itself with its response. Henry Ford II countered that the proposed safety controls on cars were "unreasonable, arbitrary and technically unfeasible. . . . If we can't meet them when they are published, we'll have to close down."[9] This comment was widely seen as an irresponsible scare tactic.

General Motors compounded its lapses in industrial design through efforts to undermine Nader personally. They hired women to lure him into a compromising position. They put private detectives on his trail in hopes of digging up embarrassing information on his personal life. This effort yielded only what Nader's associates already knew: He doesn't really have a personal life to speak of. But GM did manage to make itself and its Corvair the butt of many years' worth of cartoons and jokes about safety and cover-ups.

It was in this atmosphere of rising public disaffection with the quality of American cars and disbelief in the carmakers' concern for their customers that Fri confronted the Big Three. "You tell those guys," he said to his staff, "they have forty-eight hours to make their fuel efficiency information public. If they refuse, I'll issue a press statement announcing that they are rebuffing our request to inform the public."

In a year when gas prices were rising from about thirty-five cents to nearly a dollar a gallon, the Big Three blinked. In an instant, fuel efficiency became a staple of car advertising. Even the gas-guzzling sports utility vehicles today report their gas mileage.

By 1974, other problems had begun to occupy the agency and state officials. The federally funded road building of the 1950s and 1960s, paradoxically, had only increased traffic congestion, and with traffic came complaints about the emissions from growing numbers of cars moving at slower speeds. The Clean Air Act as passed in 1970 required the agency to come up with standards for car engines by 1975, and this deadline was fast approaching.

Lester Lave entered this debate from a wholly unexpected and largely unwelcome direction. One of the country's best young researchers in the field of econometrics—the science of turning economic behavior into understandable patterns and numbers—Lave held the rank of associate professor at Carnegie Mellon University and was working with Eugene Seskin, then an undergraduate. Lave set out to correlate two different mounds of data: those coming from U.S. counties on air pollution, and various measures of health. In 1964, the surgeon general's report on smoking and health had made the case for linking big patterns in lung and heart disease with cigarette use over large populations and long periods. Lave hoped to do the same for air pollution.

By the time EPA was up and running, health statistics had begun to come into their own. There had been no national registry of deaths until 1949, but plenty of states were using a formal system called the International Classification of Disease, or ICD, for tracking health. The development of Medicaid and later Medicare was one of the main reasons that states now needed to know more about their own populations: The states' eligibility for money depended on their providing descriptions of those in need. Out of this requirement, new bureaucracies arose for monitoring and reporting births and deaths on a national level. These reports had already provided the basis for several fascinating studies of the damaging effects of polluted air. A team of investigators from New York State led by Warren Winkelstein had found in 1967 that deaths from bronchitis were twice as high in the most polluted census tracts around Buffalo as they were elsewhere in the state.[10] In 1962 and 1964, William Haenzel and others had reported in the *Journal of the National Cancer Institute* that men and women smokers and nonsmokers who lived in urban areas had double the chance of dying from lung cancer compared with those living in rural areas.[11]

But all these findings were regarded as flawed by public health experts. Starting in 1967, in an effort to provide an improved approach to the problem, Lave and Seskin spent the better part of a decade piecing together information taken from huge black books in which the records of deaths and disease were stored. What they achieved was barely short of a revolution in public health research. It took about two decades for the public health profession to catch up.

In August 1970, *Science* published the first report from the Lave and Seskin collaboration. This came just four months after the first Earth Day demonstration and half a year after President Nixon declared his intention to address the problems of pollution. The authors began by noting that public interest in air pollution extended to the president and civic leaders and "seems to have risen faster than the level of pollution in recent years." Conceding that volcanoes, plants, and animals contribute air pollutants, they allowed that while the full impact of air pollution on health could never be completely determined, slow and steady exposures to polluted air generally could only render "living things less comfortable and less happy."[12]

Happiness is a term used with surprising frequency in economics literature. But happiness to an economist is not the same as joy to a baby. Rather it refers to an understanding that begins with Jeremy Bentham, the inventor of the *felicific calculus*—the system of counting "goods" or "benefits." Rather than try to assess something as ineffable as benefits, however, Lave and Seskin looked for ways to gauge specific things that could be directly measured. They settled on the health consequences of polluted air, arguing that the hard costs of poor health are real and at least theoretically measurable. Information was then becoming available on average annual levels of sulfur oxides and particulates in 114 standard metropolitan statistical areas across the United States as well as various U.S. counties. Lave and Seskin linked this with rates of deaths in infants and in those over age sixty-five.

To carry out this assessment, they relied on a statistical technique commonly used in economics, called multiple linear regression analysis. This method allows scientists to take large amounts of information on variables such as the death rate, annual measures of pollution, social status, and smoking and quantify the changes in their relationships. Lave and Seskin wanted to know whether differences in death rates could be explained by air pollution, income, or smoking, or by some combination of these. They looked at deaths, disease, and pollution over the single year of 1960 in all their cities, working with a cross-sectional analysis—basically a snapshot in time of all these variables at once.

The two researchers, again, were econometricians, which meant that the subject of economic measurement was never far from their minds. If the public health consequences of pollution could be quantified, then the benefits could be economically valued and the health benefits of controlling this pollution could also be calculated. In effect, economics, like public health analysis, is nothing but a sophisticated set of mathematical tools that can be applied to large sets of information arrayed in time and space. But no one had yet taken advantage of the similarity of the tools and their ability to be applied in similar ways to both kinds of data.

Health scientists of the day tended to focus their work at the level of individual well-being and had little experience in estimating costs, even for planning purposes. Economists were trained to look at the bottom line: who would pay, and how much. In a society in which cash is the only universally accepted measure of value, those who argued that pollution should be addressed just because it is wrong and distasteful remained a small, though vocal, minority. Then as now, the driving questions were, how much did it cost to keep the air dirty, and how much would it take to make it cleaner? What would be the benefit to society of cleaner air?

Lave and Seskin addressed this from a highly practical vantage point. Is air pollution, they asked, more of a public "good" or a public "bad"? In other words, do the activities that cause air pollution create more wealth than its consequences take away? It is highly doubtful that anybody from a public health background would have had the cheek to take this on. It was pretty cheeky even for economists: The standard practice in economics was to regard any activities that result from pollution—whether cleanup costs, hospital stays, or burial costs—in terms of their added value to the gross domestic product (GDP). This activity was generally seen as adding to GDP, and therefore pollution was a "good." By creating a system of accounting for the negative effects of pollution on health, Lave and Seskin started what would ultimately become *green accounting*—a way of tabulating the positive and negative economic consequences of various policies beyond simple GDP values.

The two authors pointed out that previous efforts to link pollution and health had looked at only one set of relationships at a time. Alfred Hexter and John Goldsmith, for instance, had studied carbon monoxide levels and death rates in Los Angeles over a twelve-month period. But in the real world, exposure always involves more than one pollutant at a time. In their *Science* article, Lave and Seskin noted that no one had yet asked about the "long-term effects of growing up, and living in, a polluted atmosphere."

In this remarkably important and typically restrained paper, the two economists laid down a framework for describing connections between health and air pollution that has not been basically altered for more than three decades. Their language was dry, their findings anything but. They showed what many had suspected: People living in areas with the dirtiest air had the highest risk of dying. The poorest and oldest nonwhite populations, living in the most polluted areas, had the poorest health of all. Men and women showed almost the same effects. Lave and Seskin concluded that a 50 percent reduction in air pollution would reduce the costs of pollution-related illnesses, days of work lost, and early deaths by at least $2 billion each year, amounting to 4.5 percent of all health-care costs at the time. They also noted that the full brunt of pollution extended far beyond the monetary costs that could be calculated.

Lave and Seskin knew at the start that whatever they did would be challenged. Looking first at what society was then paying for medical costs to treat people for illnesses tied with air pollution, they then added the value of earnings lost by those who died or became sick. This technique, they declared, yielded a "gross underestimate." Yet it showed that in 1970 the United States paid nearly $1 billion a year in lost work, deaths, and bronchitis treatment alone. They had not, they admitted, taken into account such things as the impact of carbon monoxide on the nervous system and heart. There was not enough information to assign these a number.

Some thirty years later, reading the article is an eerie experience. Lave and Seskin dismissed the views of those who believed that only high-pollution episodes caused health problems: "It is the minimum level of air pollution that is important, not the occasional peaks. People dealing with this problem should worry about abating air pollution at all times, instead of confining their concern to increased pollution during inversions."

As a "gross underestimate," $1 billion a year was hard to accept, both because of the amount and because of how the estimate was obtained. Many of those trained in public health felt certain that if air pollution were really causing so many deaths and illnesses, then doctors would already have figured this out. Doctors had known for centuries that people who worked with hazardous agents paid a price. Bernardino Ramazzini, in the sixteenth century, had documented debilitating illnesses in those working in the "dangerous trades"—including printers, chimney sweeps, painters, and builders. Engels and Dickens had depicted the deadly workplaces forced on children and other workers in nineteenth-century England. Physicians in industrial medicine were well schooled in treating lung injuries from high exposures to corrosive or allergenic compounds. But the notion that millions of records on deaths and disease could tell us anything important about pollution struck many as highly questionable. It did not help that Lave and Seskin were economists.

Here was the paradox of this work. How could patterns of death and disability throughout the United States be understood to reflect the physical environment, when doctors already knew that so many things that determine health were highly individualized? People go to doctors one at a time. Physicians diagnose those who are ill, testing lungs, measuring pulses or blood pressure, drawing blood, and recommending remedies. Meanwhile, some health scientists launched costly and detailed medical studies that intensively compared a few hundred persons living in a pristine area, such as Chilliwack, British Columbia, with those living in regions with a very different pollution profile, such as Berlin, New Hampshire. No differences were found in the lungs of subjects in these two distinct regions. Therefore, reasoned the health scientists who conducted these studies, air pollution was not really that important.

This dismissal by public health experts of Lave and Seskin's work was a variant of something quite common in the history of science. Especially when they come from outside a particular field, new ideas are often dismissed as being either incorrect or unimportant. Sometimes they are deemed both. Later, these new concepts may become accepted as possibly correct, especially once the appropriate field discovers them independently, but then they are often deemed not that important. Breakthroughs in science are finally embraced when those who denied them in the first place claim to have actually made the original discovery all along.

This pattern is pretty much what has happened with the work of Lave and Seskin. But something more sinister was going on at the same time. Not only did the men have to fend off the public health community's resentment that two economists had dared to tackle health issues and done so in an interesting and novel way. Because many health researchers took issue with the audacity of the economists' work, it was not that hard to find someone who would accept payment to discover deficiencies in Lave and Seskin's research. Lave remembered that the Electric Power Research Institute hired Ron Wyzga directly from Harvard's School of Public Health to find problems in the work of Lave and Seskin and any others who weighed in on the subject. Wyzga contracted with another Harvard statistician, Larry Thibideau, to redo Lave and Seskin's analysis top to bottom. He spent several years in the effort only to confirm their basic findings. Others were not so honest, either about who paid them or what they found. The young professor Lave and his even younger student collaborator found themselves subject to a barrage of criticism.

Lave recalled that the storm of interest in this work went on for some time, as he was invited to give many public lectures on the subject:

> Some people tried to get me to be more alarmist but most insisted that the association was not causal and that they could not accept the results because we did not control for smoking or personal habits. They pointed out that people are only outside for a couple hours per day and that indoor air quality is different from outdoor air quality. They pointed to the inaccuracy of the air monitors and to the analytical techniques to get sulfates.
>
> The net result was that the public wanted to believe our results and the public health people didn't. I remember an interview with the editor of the newsletter of the American Public Health Association. He gave us

a headline in the newsletter but hastened to tell me that he didn't believe any of our work.

Those running EPA in its early days did not quite know what to do with the early work of Lave and Seskin, or with other findings that were beginning to support their views. Del Barth, then working in the National Air Pollution Control Administration, had presented a paper to the annual summer meeting of the Air Pollution Control Association in 1970. He calculated that if the maximum allowable levels of air pollution from all sources could be lowered about 90 percent, the result would be a real improvement to public health.[13]

The Clean Air Act had been under consideration all summer long. Bill Pedersen, one of the agency's young staff attorneys in 1972, recalled that Ed Tuerk, who ran the air pollution control program at EPA, told him how these numbers got put into law. One day in 1970, a call came in from a Senate Environment and Public Works Committee staff member, Leon Billings, asking the agency to weigh in on the public health dangers of air pollution. "Look," Billings said, "we can pass a bill requiring a 90 percent reduction in air pollution from cars over 1970 levels if you can tell us it's bad for public health."

The agency faced many quandaries, but here they could point to published research. In fact, the Barth study dealt with what would happen to public health if air pollution from all sources were reduced 90 percent. But it fit the bill close enough that the figure was put into the legislation, requiring that cars lower their emissions by this amount.

The question Billings raised brought home to many the extent to which EPA in those early days was flying blind. If published research was either immediately attacked as highly controversial (like Lave and Seskin's) or did not exist, then how was the agency supposed to make decisions on what to do?

One of the first studies ever begun by EPA was the Community Health and Environmental Surveillance System (CHESS), pulled together from leftover research programs begun by the Department of Health, Education, and Welfare's office on air pollution. CHESS was supposed to take a snapshot of pollution levels and health conditions in a variety of communities across the United States. It was begun on a shoestring budget, with the thought that more money could always be found later.

Making the leap from Department of Health, Education, and Welfare to EPA was a young physician named Jack Finklea. He would later head up the National Institute for Occupational Safety and Health and serve as assistant surgeon general of the United States. In those days, his team had an almost gleeful anticipation of their work. Many of the team's members had seen firsthand what had happened in Donora, and they had been on call when London's killer smog hit. They had begun to track reports of repeated, but smaller episodes from many cities. Well aware of the groundbreaking work of Lave and Seskin, they leapt at the opportunity that CHESS seemed to present.

Seeking to understand the full range of impacts of air pollution and having no serious funding to do so, Finklea and his team scavenged for any available information they could find and began speaking to anyone who would listen about their hopes for air pollution research. In addition to Lave and Seskin's troubling findings from 1970, there were other incriminating studies. One study showed that rates of bronchitis and other lung disease were higher in more polluted cities at the time, such as Pittsburgh and Chattanooga. An autopsy study of young men who had died in accidents found higher levels of carbon in the lungs of those from St. Louis compared with those from Winnipeg, Canada. An earlier report from CHESS had concluded that although cigarette smoking played a major role, air pollution in itself remained a very big deal and was "a significant, independent and consistent contributing factor" for bronchitis and a host of other lung diseases.

Having begun to link all the current, available pollution and health information into geographically useful categories, the CHESS researchers talked up their work to anyone who would listen, including members of the New York and California congressional delegations. Word of a big EPA study about the nationwide effects of pollution filtered back to the White House. In 1999, Finklea told me that in 1971, the Office of Management and Budget (OMB) called him in to explain the request for major funding for CHESS in 1972.

For a federal official, a summons to OMB, no matter what the stripes of the administration, resembles nothing so much as a trip to the woodshed. OMB remains the cop at the door with respect to which agencies get how much money, when they get it, and for what. Its job then as now is basically to keep costs low and output high, with the emphasis greatly on the former. Nixon's OMB was especially powerful; he had ordered it to "impound" (i.e., refuse to spend) money appropriated by Congress for certain purposes deemed too liberal.

In preparation for his session with OMB, Finklea brought along the usual charts and data, ready to defend the work and most importantly its costs on scientific grounds. Up to that point, CHESS had relied on some rather dubious sources of information. For instance, air pollution monitors for sulfur oxides and carbon monoxide had been located right atop smokestacks to measure the effectiveness of pollution controls. This did not provide a good indication of what people actually breathed. Finklea's team needed to see that monitors were set up in more appropriate places, ensure that they used the same methods for measuring the same pollutants, and organize the monitoring process so that measurements in different regions could be compared with one another.

The OMB examiner whom Finklea met with was a fellow named James Tozzi. Tozzi later became associate director for Natural Resources, founded a successful lobbying firm with several Fortune 50 companies among its clients, produced a limited edition of his own bottled cabernet sauvignon from France, and owned a trendy restaurant in Washington's Dupont Circle area. At this encounter, Tozzi did not seem especially interested in the details of the science. He seemed rather more concerned with what Finklea might have told members of Congress.

Finklea remembered beginning, "Mr. Tozzi, I can explain to you exactly what we're doing, how we're going about setting this study up, meeting all the important scientific requirements. We really need this budget so that we can get better quality information on exposure. We just do not have good enough data on what people are being exposed to."

Tozzi cut him off. "Dr. Finklea, I'm sure you can do that. I don't doubt that you could use more money to get better information. In fact, that's precisely the problem. I think the Congress does not need to hear all these details. I don't think the country is ready for that."

It was clear that the clock would soon run out on CHESS. If Finklea's team did not come up with better-quality data, they were doomed. Within two years, just as he feared, the CHESS studies ended. If you don't want to know the answer, just don't ask the question.

For CHESS, as for Donora, no final report ever appeared. If it had, it surely would have mentioned instances of lethal smogs in New York, Birmingham, Pittsburgh, and many other industrial centers—deadly episodes that were recognized only after statisticians went back and looked for them. So we did not learn until years later that in 1953, 1963, and 1966, hundreds more deaths than usual occurred in New York City. Over Thanksgiving weekend in 1966, the entire northeastern United States sat under an unusually warm and unhealthy high-pressure system for several days. In New York City, 24 deaths more than usual occurred every day for at least five days, or 120 more deaths than would otherwise have happened.[14]

These episodes of air pollution are detected only by statistical analysis. Especially in very large cities, where a hundred or more will normally die every day in any case, a few dozen deaths more than usual are impossible to notice. Someone has to care enough, and be given the opportunity, to look in the first place and be expert enough to find a correlation to some plausible cause. Then that person has to have the resources to get the information out, once it is found. The CHESS study failed on the first count: The opportunity to do the research was denied. Lave and Seskin nearly failed on the second.

In the academic world, research tends toward two extremes. There are those, like lawyers, for whom footnotes and references fill more space than the text itself. Others emulate theoretical physicists, for whom too many references signal work that is derivative rather than original. Encroaching on a foreign field and needing to establish their bona fides, Lave and Seskin wrote like lawyers: they set out to show that they had taken into account every possible alternative explanation for what they had found. Seven years after the first major paper in *Science* and a decade after they had started their research, they published a 368-page book with more than 100 pages of references and appendices. This exhaustive work documented with unprecedented thoroughness that air pollution caused serious public health problems requiring major changes in social policies. The authors had spent the better part of the 1970s in a variety of obligatory seminars before the high priests of statistics and epidemiology, defending their approach.

The story of the book's publication itself provides a good case study of how the necessary conservatism and rigorous skepticism of science are sometimes hard to distinguish from obstructionism. The book was originally scheduled to be published some years earlier than 1977. In 1974, Lester Lave was summoned to the august offices of John Tukey, then the chairman of statistics at Princeton University. "I was flabbergasted," Lave recalled. "Professor Tukey was a very important, very intimidating man. I had some claim to being an econometrician, but none whatsoever to being a statistician. I was certainly outranked, a measly economist in the den of statisticians."

The Princeton Statistics Department had split from Mathematics just a few years before, in 1966. The young department—the best in the United States—contained a roster of faculty members whose names had been given to various widely used tests in statistics, such as Fisher, who had devised Fisher's exact *t*-test of statistical significance.

Lave had sent Tukey the galley proofs of the book, and Tukey had gone through it meticulously, making suggestions for changes and pointing out problems. His advice was powerful: This is too important an issue to publish a book with these problems. During a seminar and workshop, the statistical issues got a thorough airing. Error after error, misstatement after misstatement, was exposed to humiliating light. By the end of the day, Tukey's position was clear: Lave needed to start all over again, recollect the data, and redo all the estimations, taking care to satisfy statisticians as well as econometricians. The long-awaited publication that was supposed to vindicate Lave and Seskin's work and answer their critics would have to be indefinitely postponed.

It's impossible at this remove to know what Tukey's motives may have been. Perhaps he had hoped that Lave would be so cowed by the critique that he would never surface again. But it would have been far more devastating to Lave if Tukey had let the book be published with its flaws intact and then had written a killing review in some prominent journal. I prefer to think that, in his severe way, Tukey was acting on the side of the angels and offering Lave a chance to save his scientific soul. Lave took it.

He and Seskin spent another year collecting data all over again, redoing analyses again and again, addressing each of Tukey's criticisms one by one. Finally, having made repeated confirmations of their findings, Lave and Seskin in 1977 issued their massively documented analyses under the title *Air Pollution and Human Health.*[15] In retrospect, Lave said that "Tukey's comments were the best thing that could have happened. It was a disappointment to spend another year reworking the analysis, but Eugene Seskin and I felt that it was worth it to increase its scientific validity."

The book began by conceding that the only unequivocal measurement of health status is death and then offered a litany of caveats. "If death is unambiguous, mortality rates are not. Death rate statistics based on cause of death are no better than the accuracy with which the cause of death is determined." The authors went on to list the uncertainties that could affect death statistics. They pointed out that even autopsies make mistakes. Because people move, their residence when they die does not necessarily tell you where they spent their time living and breathing. Measurement of pollution is also less precise than anyone would have liked, the weather has not been well studied or measured. They admitted the limits of all sorts of information. But in the end, they held up their conclusions of nearly a decade earlier.

This book did something that their 1970 article had not. By amassing so-called time-series information, Lave and Seskin looked at how pollution and health changed over time. Time-series information tracks what happens over years as populations age and pollutants change. No matter how the authors crunched the numbers, the connection remained. Air pollution kills and sickens people, especially the very young and old. And the dirtier the air and the poorer and older the people, the greater the impact pollution has.

In summing up their reasons for concluding that air pollution directly caused poor health, Lave and Seskin relied on a way of thinking that had been used by the Royal College of Physicians in England in 1962 and by the U.S. surgeon general's report on smoking and health in 1964, both of which determined that cigarette smoking caused lung cancer and heart disease. Keep in mind that at that point, there was no definitive proof of how cigarette smoke might induce either disease.

Human studies had consistently found that those who smoked the most had the greatest risk of the disease. But much remained unresolved. These reports had relied on what became known as "the full weight of the evidence" available to them, ranging from experimental findings to clinical and epidemiologic studies to basic biology. The Royal College and the surgeon general had emphatically not insisted on definitive proof of human harm before concluding that smoking was hazardous to human health.

Later, Sir Austin Bradford Hill, one of England's foremost environmental health researchers, had come up with what he termed several "logical viewpoints" that could be employed in deciding whether causation could be said to exist for any given situation. How strong was the association between the suspected exposure and the diseases? How consistent were the findings? Did those with less exposure have less damage? Did an end to the exposure reduce the damage? Does the whole thing make biological sense? Are there experimental data that support the association?

Hill was clear on one thing. No hard and fast rule exists for making a determination of causation. But the reasoning laid out by the Royal College and the surgeon general allowed health researchers to ask one fundamental question: Is there any other way of explaining the set of facts before us, is there any other answer equally likely, or more likely, than cause and effect?

Here is where Lave and Seskin's book really turned things upside down. Never again would an EPA official assert that evidence on the hazards of air pollution was lacking, nor would an industry representative get away with arguing that human harm remained unproven. The authors conceded that correlations between any two things, even things that seemed intuitively connected like bad air and lousy health, did not necessarily show that these two things were in fact directly linked. They had spent the better part of the past decade extending their regression analyses to look into most of the so-called confounders, such as smoking and poverty, and showed that these could not account for what they kept finding. By 1977, they had information on deaths and air pollution for the entire decade of the 1960s. Even in areas for which the rates of smoking were similar, people who lived where the air was dirtier throughout the decade still had higher rates of illness and death. Not only had they demonstrated, in far greater detail than ever before, the statistical correlation between air pollution and bad health, but they also emphatically closed the door on there being any more likely explanation than cause and effect.

Taking an even finer look into all this, they pulled together deaths from every single day for five years in Washington, D.C.; Chicago; Denver; Philadelphia; and Saint Louis, from 1962 to 1966, and assessed these against air pollutants measured at the same time. Using statistical techniques that allowed them to see beyond correlations to real relationships between many different variables, they even incorporated a lag period into their system. Health on a given day does not just tell you about the air breathed on that day, but also indicates the results from air breathed for several days earlier. Lave and Seskin reasoned that having several consecutive days of higher pollution would be worse for health than just having occasional bad-air days.

Only in Chicago, where pollutant levels were ten times those in Denver and four times those in Saint Louis, did they find a significant link between death rates and pollution. This was not because the other cities were free of problems, but because their study in these other cities lacked what is called statistical power to find an effect. To find a problem in a given population, you have to study a large enough group, or sample, of people. Moreover, the factor you are studying has to be either very common or very rare in order for it to show up. For things that are very rare, you must study correspondingly large numbers of people. Large samples are also required for things that are very common, because these things will occur naturally and you need enough cases to show some anomaly. Because death is not just common but universal, studying connections between pollution and the death rate required examining records on millions of people over long periods. In this regard, power is to statisticians what a flashlight is to an explorer of caves. Without sufficient light, you are left to use your hands to feel around a cave, but you cannot see the cracks and crevices with sufficient detail. Power effectively gives statisticians the ability to see things they otherwise could not.

In a city like Chicago in 1970, with a population of more than 3 million, each day saw about 81 deaths. By looking at patterns in the numbers of people dying over a year, we can calculate the amount that a rate will vary, known as standard deviation. For Chicago, the standard deviation is about 15. This meant that most of the time, the number of deaths each day in Chicago would fall between 96 and 65. Most of the other cities in Lave and Seskin's group had populations of half a million or less. On an average day, about 13 deaths would occur in those cities, the normal range being anywhere between 21 and 5. Thus, you need a much greater percentage of change in the death rate before you can notice a difference. Studies in larger cities will obviously give you a better chance of finding patterns than those that look at smaller areas.

These results were of stunning importance for the United States. Chicago may have been unique in many ways, but not in how its citizens breathed. Finding effects in this city meant that similar effects were occurring everywhere else.

In 1973, Bob Fri did not know all the controversy that swirled around the work of Lave and Seskin. What he did know was that air pollution was growing wherever it was being measured and he was in charge of the agency that had to develop standards for cars and gas by 1975. People were alarmed and dismayed about visible pollution and were fleeing the polluted zones of the industrial Northeast in growing numbers. Complaints about health in polluted zones were landing on Fri's desk, and the law required him to act to reduce that pollution. He also knew that some of the car companies were doing what they had done for years—stalling.

Armed with some numbers from Lave and Seskin's work and with piles of reports from its own engineers, EPA set about to do what Congress had ordered, namely, get cars under control. The car companies were giving the agency a hard time. Throughout the 1970s, whether speaking to regulators, Congress, or the public, auto makers consistently held that any effort to impose federal engineering requirements on car engines would prove their undoing. Some of their representatives, Fri remembers, "just lied straight out. We knew that they could comply with what we were asking. They told us they would go bankrupt. This was nonsense."

On the matter of making public how far their cars could travel on a tank of gas, Fri got the response he wanted. Within weeks, the same firms that had sworn they would be ruined by publicly available information on fuel efficiency had devised marketing campaigns trumpeting their mileage figures. Fuel efficiency found its way into advertising long before it was written into law.

Next on Fri's agenda was getting the car companies to reduce what was coming out of their tailpipes. He did not have the luxury of waiting for health studies to be done and redone. The law creating EPA stipulated that car emissions had to be brought down 90 percent below levels allowed in 1970 for hydrocarbons and carbon monoxide by 1975. As to how the reductions were to be accomplished, that was left up to the manufacturers. The reaction, not surprisingly, was nothing short of horror.

It was well known that the efficiency of modern car engines depends on how fully they burn fuel. The most efficient burn rate is produced when the ratio of air to fuel is near the ideal—what is called the stoichiometric point. In theory, at this point, one can minimize emissions of hydrocarbons, carbon monoxide, and nitrogen oxides, using all available oxygen. For gasoline, the stoichiometric ratio is about 14.7 to 1, meaning that for each pound of gasoline, 14.7 pounds of air are burned. Not surprisingly, driving conditions and temperature and pressure changes within an engine can alter this ratio. As a result, the amount and types of compounds released from tailpipes can vary a great deal. Because internal combustion occurs under varying conditions, nitrogen—the most common gas in the atmosphere—combines with oxygen in a number of ways, producing oxides of nitrogen that can be highly reactive and highly toxic, along with unburned residues of gasoline.

The composition of the main tailpipe emissions of a car are among the most commonly occurring materials on the planet—nitrogen oxides (NOx), carbon dioxide (CO_2), and water vapor (H_2O). This chemistry has not changed for decades. An advertisement from 1928 boasted that an automobile is nothing but a chemical factory on wheels. Every gallon of gasoline consists of about 1 pound of hydrogen and slightly more than 5 pounds of carbon. When burned, these yield water, oxygen, carbon monoxide, and carbon dioxide and nitrogen compounds. In addition to these widely occurring compounds, the car spits out small amounts of incompletely burned gases, including a laundry list of toxic and volatile materials such as benzene, butadiene, n-hexane, and other highly aromatic compounds that give gasoline its distinctive odor.

The representatives from Chrysler took a hard line: Changing engine assembly so quickly would put them out of business. It was not simply a matter of sticking a new device into the tailpipes of cars. Many other engine specifications would have to change—and this, said the Chrysler officials, could not be done quickly or cheaply. A senior agency official engaged in the negotiations recalled, "The fellow from Chrysler came in with a whole group of experts, tables, charts, and all the rest. His view was this was absolutely not going to happen on his watch. He feared his company could not retool fast enough. [The Chrysler man told us] 'We cannot do this. This will bankrupt us.'"

Foreign rivalries were already pretty contentious. In response to reports that Honda had created an engine that met the proposed U.S. 1975 standards in 1973, one Senate witness from GM reportedly said, "Oh sure, they can do that with their little bitty Japanese car, but not with a real big American car."

This slap did not really sit well with Soichiro Honda. He bought a Chevrolet Impala, shipped it back to Japan, and had his engineers replace the existing engine cylinder head to prove that their system could handle bigger, heavier cars. Eric Stork took the car for a test spin that summer in Japan. A few months later, addressing a big event sponsored by GM, Stork reported on his test drive of a large American car fitted with a working Japanese pollution control system. He began his speech with this story:

> Blue Suit Burns was an Irishman with lots of friends, who spent a bit too much time in bars. He, sadly, died young from cirrhosis of the liver. Before the funeral, Mrs. Burns went back to see her husband and found him laid out in a brand new brown suit. She became distraught when she saw him in the wrong-colored suit and began to yell at the undertaker.
>
> Much to her surprise, at the funeral, there was Blue Suit Burns dressed as usual in his customary color. Mrs. Burns began to apologize profusely to the undertaker.
>
> "I am so sorry for having yelled at you."
>
> The undertaker replied, "Oh, please don't bother yourself."

Mrs. Burns insisted, "I am really sorry, I was so upset."

The undertaker replied, "Really, it was not a problem."

Mrs. Burns again apologized. The undertaker explained, "Really, Mrs. Burns, it wasn't any trouble at all—all we had to do was to change heads."

Changing heads of either cylinders or company executives was not a simple matter for the car manufacturers. It was the new fuels, Chrysler now claimed, that would be their undoing. Gasoline could contain high amounts of sulfur, which influenced emissions and also could damage the catalysts in catalytic converters. Gasoline consists of carbon and hydrogen chains of different lengths, ranging from C_7H_{16} (seven carbons) through $C_{11}H_{24}$ (eleven carbons). Because gasoline never does burn completely, lots of other by-products come along whenever it is used in a car engine. Chrysler insisted that the oil refineries would not be able to provide the cleaner-burning fuels needed and that the company could not build the required new engines. It had bet on the wrong technology—something called lean burning, which depended on improving how fuels were formulated.

General Motors, on the other hand, had come up with a much cheaper approach to the problem. A remarkably short time after the Clean Air Act was passed, GM developed a kind of chemical washing-machine device that could be added to a car's exhaust system. There, the device would chemically break apart many toxins coming out of the engine. Toxic emissions from the engine, such as nitrogen dioxide, would be broken into its two elements. The device would convert carbon monoxide by combining it with another oxygen molecule to yield carbon dioxide and water. This device was called a catalytic converter, because it worked by changing the chemistry and composition of the substances that finally left the engine.

But there were major problems with this new device, the company assured the agency. There was no way that companies could be expected to reliably produce catalytic converters for the mass market any time soon. Cars would become unreliable, and companies would flounder. Here is what Earnest Starkman, a vice president of General Motors, testified to the Senate Committee reviewing the proposed regulations at a public hearing in 1972:

> If GM is forced to introduce catalytic converter systems across-the-board on 1975 models, the prospect of an unreasonable risk of business catastrophe and massive difficulties with these vehicles in the hands of the public must be faced. It is conceivable that complete stoppage of the entire production could occur, with the obvious tremendous loss to the company, shareholders, employees, suppliers, and communities. Short of that ultimate risk, there is distinct possibility of varying degrees of interruption, with sizable dislocations.[16]

Starkman came to the EPA offices in 1973 and went through a long explanation of all the major hardships that converters would produce for their company. By then, however, the company understood that it was not a question of whether all this would happen, but when. "Ed Cole, then an executive with General Motors, left us the briefing books his folks were using to lobby Capitol Hill," said Fri. "He was quite candid. 'Look, we can do this. We can meet this standard now, for most of our cars. Problem is the timing is just wrong. We need another year to ramp up our production.'"

A team from Ford also came in for a lengthy briefing and was far less cooperative. They explained how complicated the new emission requirements were and how hard it would be to create such engines on a large scale. Bill Pedersen recalled years later the impressive swath the lobbyists cut: "All the Ford team had been well coached by their lawyers. In their arguments on why we should not do this, they made EPA pay for every defect in our proposals—and there were plenty of them."

In a speech to the Chamber of Commerce in New York on February 15, 1973, Ford's president, Lee Iacocca, warned of doomsday for the auto industry:

> We could be just around the corner from a complete shutdown of the U.S. auto industry and all that that implies. . . . Baked into law is a requirement that by 1975—or 1976 at the latest—we reduce the three pollutants that come [out] of our auto exhaust pipes— hydrocarbons, carbon monoxide, and oxides of nitrogen—by 90 percent from the levels of 1970 and 1971. We've already spent millions of dollars and we can come

> pretty close, but no cigar. We can't meet the precise requirements of the law. And that's the hang-up.
>
> For example, Ford's 1973 models have hydrocarbon levels that are 85 percent less than those produced by cars without emission controls a decade ago. Our industry has been backed to the cliff edge of desperation, and time is running out. . . . As far as the even-tougher 1976 requirements are concerned, nothing seems to describe our situation as well as that clincher by Sam Goldwyn "In two words: IM . . . POSSIBLE!"[17]

We would do well to remember this speech the next time someone like Senator Trent Lott holds up a photograph of a Mini Cooper and threatens that if car companies are forced to raise fuel efficiency by 3 miles per gallon, this is what we'll all be driving. What was nonsense a generation ago is still nonsense today. While Iacocca's bold declaration made for impressive drama, it ignored the reality that Honda and Mazda had already developed cars that met the new U.S. standards. By January 1974, EPA knew that it was holding the trump card when Iacocca sent his team off after they had gone through a detailed briefing and began arguing that his company could comply, if it was given a bit more time to do so.

EPA couldn't require catalytic converters in cars—only that cars be in compliance with emission standards. Still, converters seemed to be the most promising way of meeting the standards. In this round of last-minute lobbying, the car companies had shifted their position, from insisting that they could not comply with the regulations to arguing for more time. Time, of course, is money. But time also meant more pollution for more people. In fact, William Ruckelshaus in 1973 had issued a ruling giving the companies an extra year to comply with the law. Then something of a technology mystery happened. Just six weeks after it had won an extra year's delay based on technical arguments that it would be unable to comply with the standards, General Motors announced that it had found a stunning, new breakthrough technology that would allow the company to meet the new standard for all cars in both California and the rest of the United States that very year. And, oh yes, it also happened to improve fuel efficiency. Miraculously, the catalytic converter could be made to work for the mass market after all, and GM just happened to have the best technology around for doing so.

Stork had been proud of the brilliant, young engineers who had long ago figured out that catalytic converters would not only reduce pollution but improve fuel efficiency. He remembers being pleased when GM publicly announced what the agency's engineers had known all along. As he told the *Detroit Free Press* at the time, "It's lonely out there telling your boss that such devices will work when everyone else is saying otherwise."[18]

For those working on the issue at EPA, the litany from the car companies became like one of those operatic refrains that gets repeated so much you can't quite forget it, no matter how hard you might like to try. "Trust us, there really is no problem. And by the way, what you are asking for, really can't be done. But now that we have figured out a way to fix it, would you just go away and leave us alone."

When I visited Lester Lave in 2002, he was not bitter over the reception accorded his work in the 1960s and 1970s. He had achieved many distinctions in his long career. But he had not done any follow-up research on air pollution in the United States for many years. He told me about how annoyed the epidemiologists were with his findings. At first, he kept getting invited to intellectual shoot-outs at schools of public health, where he would be peppered with questions about lung physiology, on which he never pretended to be an expert. All he knew was that his work made clear that what Goldsmith and Breslow had shown in Los Angeles in the 1950s and 1960s was also happening across the entire country. Gradually the invitations stopped coming, and he moved on to other arenas. When I tried to locate a copy of this seminal book at Carnegie Mellon's library or at the University of Pittsburgh, none could be found. Lave himself had one.

There are other copies sitting on library shelves, but nobody reads them. Lave was too far ahead of his time. The world was not ready to accept the implications of his work, and the pressures to keep things going as they were proved far more powerful. Those captains of industry who delayed adopting cleaner cars and energy plants probably never lost a night's sleep over the thought that they had within their power the means of protecting millions of lives.

All the opposition, honest and otherwise, had a chilling effect on research in the field for years. Resistance to this work, which went far beyond the normal bounds of peer review, kept the heat turned up so high that for quite a long time, few students or researchers entered the field of air pollution epidemiology. Science does not reward those who take on controversy—at least, not while they are alive. Funds for such work were scarce. If you don't want to know, don't ask the question.

In 1980, several health scientists from Harvard, including Jack Spengler and his brilliant mentor, the nuclear physicist Richard Wilson, confirmed and refined the basic findings of Lave and Seskin and created another monumental record of evidence that low levels of air pollution constituted a major threat. Their book *Health Effects of Fossil Fuel Burning* not only presented health information but provided for large-scale analyses of wind, weather, pollution, and health, along with detailed engineering descriptions of various forms of electricity generation and transportation.[19] The findings of Lave and Seskin had not changed. If anything, they became more robust—more solid, less vulnerable to changes in the data, and less easy to dismiss. Tapping work under way at Brookhaven National Laboratories by Leonard Hamilton and others, Spengler and his colleagues made headlines when they estimated that 50,000 people died from air pollution in the United States in 1980.

The scientists directly attacked a long-held "truth" in public health, namely, that there are levels below which pollution has no biological impact. These levels were referred to as thresholds and were thought to indicate amounts of exposure that were safe.

The authors explained that biologists typically work at the macroscopic level of the whole person and draw upon evidence that the body can usually survive or detoxify itself, providing that exposures are not too high at any point. But coming from the field of physics and engineering, Spengler and his colleagues took a more microscopic view. They examined molecular interactions that could occur in many different components of a living system, from the cilia lining the upper breathing zones to the invisibly small regions of the lower lung. Some damage, such as reduced growth or acid injury to the lung's linings, could not be repaired, even if individuals who experienced it showed no overt evidence of illness at the time. Eventually, these insults would make those who endured them more vulnerable to a host of other health problems.

In addition to looking solely at such microscopic exchanges in a single person, Spengler and his associates looked at what such damage would mean for hundreds of thousands of people, much as Lave and Seskin had done. They began to identify large patterns of disease and death that were tied to air pollution and that could be prevented. A chilling thought underlay their work: What if air pollution did not have a threshold?

For Spengler and his co-authors' book, released a decade after Lave and Seskin's groundbreaking paper in *Science*, the same cycle of criticism ensued, fueled again by economic forces. "I remember our book had just come out," Spengler told me. "One of my medical colleagues—far senior to me—dropped by. He looked at my new book and asked, 'Mind if I borrow a copy? I'm being paid to criticize it.' That was my introduction to how the game is played."

Part Two

THE BEST OF INTENTIONS

In centuries past, political intrusion into science has led to horrifying tragedies. Antoine Lauren Lavoisier laid the foundations for understanding chemical elements in the eighteenth century. A member of the French National Academy from the age of twenty-five, Lavoisier set up an international center for chemistry where scientists from throughout Europe gathered to exchange the latest developments. When the Jacobians took over France, the cosmopolitan and aristocratic Lavoisier immediately came under suspicion.

Lavoisier, the revolutionaries alleged, had committed a crime against science. In 1794 this seminal thinker, the discoverer of oxygen, was charged with the dual offenses of internationalism and of diluting tobacco with water. He was imprisoned together with his wife. His crime was endangering public welfare. Her crime was to be married to him. He was guillotined on May 8, 1794, at the age of fifty-one. A few days afterward, Joseph-Louis Lagrange, a distinguished mathematician who had not come to the attention of the revolutionaries, remarked, "It took only an instant to cut off that head, and a hundred years may not produce another like it."[1]

Although the guillotine has since been retired as a tool of scientific discipline, scientists remain vulnerable to other pressures. Who decides what information should be looked at in order to determine whether a given compound or process should be considered a hazard? What types of evidence are most important? How are statistical tests to be applied? Who controls what research gets funded, how it gets looked at, by whom, and how much money will be spent to gather new information? Who decides which experts will be asked to review proposals and papers? All these questions are not just scientific matters. They can affect the health and prosperity of millions.

5

ZONES OF INCOMPREHENSION

> The genius of these evils was their ability to create zones of incomprehension.
>
> —SAUL BELLOW

IN THE IDEAL WORLD, science works only if it remains objective, independent and outside of politics. While the decision about what to study or even whether to study anything at all may legitimately come from the personal values of the individual scientist, the methods applied to the research effort are supposed to be isolated from those values. Long before the Nazis banned "Jewish" physics and Soviet hacks tried to make biology conform to dialectical materialism, the German social theorist Max Weber warned of the dire consequences of allowing politics to dictate scientific results.[1] More recently, the feminist philosopher of science Sandra Harding has explained that what passes for scientific rationality changes throughout history because of the "constant reinterpretation of what should count as legitimate objects and processes of scientific research."[2]

To maintain some claim to objectivity, science today awards fellowships for students, grants money for research, and publishes papers under a formal process called peer review, which relies on anonymous evaluations of proposed research and writing. Criticism, sought from experts or peers most qualified to give an assessment, is offered without concern for offense or personal relationships. The whole process is supposed to work democratically because it remains insulated from political and economic pressures. That, at least, is the theory.

But democracy can create other problems for science, especially when laws are passed requiring technologies that do not yet exist. The Clean Air Act of 1970 can be thought of as a basic science and technology forcing law that, among other things, obliged EPA to issue standards for reducing the use of lead in gasoline and a timetable for ultimately phasing it out altogether. Lead residues in dust had been found to track neatly with road traffic, with higher levels falling closer to major thruways.[3] With the Arab oil embargo in 1974, EPA's plans to phase out the production of lead were put on hold. It was argued that the increased energy required to remove this heavy metal from gasoline would reduce the oil supply at a time when the United States was still reeling from the spectacle of people standing in line for hours to buy fuel.

With its track record of successfully opposing all past efforts to reduce lead in fuels, the Ethyl Corporation sought to capitalize on the oil embargo and made a renewed effort to overturn EPA's regulations when they resurfaced in the early 1980s. The company premised its position on the assumption that until and unless the hazardous nature of a substance had been proven to afflict humans, the material should be used where the known economic benefits were believed greater than any potential harm. From the vantage point of the company, therefore, it was important to make sure the dangers from putting lead into fuel remained uncertain. Throughout its history Ethyl tapped its own scientists and hired experts from universities to argue that levels found in the environment from gasoline were trivial for human health and to challenge any studies suggesting that low levels of lead were hazardous.

The most convincing analysis of the dangers of lead was a profoundly disturbing study published in the *New England Journal of Medicine* in 1979 by Herbert Needleman, a toxicologist and psychiatrist then at Harvard University.[4] The amount of lead in the blood, he wrote, only tells you about exposures that have happened in the past few months. Lead eventually ends up in bone or brain, where it cannot easily be measured. Needleman reckoned that baby teeth, with their soft dentine centers through which blood circulates, would integrate the total amount of lead to which a young child had been exposed. Looking at residues of lead in these discarded teeth collected over many years, he showed that fifty-eight elementary-school children in Massachusetts with "high" lead levels performed consistently less well on a variety of tests of intelligence and behavior, compared with one hundred children with "low" levels. Those who had higher lead levels as toddlers had IQs nearly four points lower on average by the time they reached the age of ten. Needleman's work meant that for the entire country, millions of children would be just a little bit less smart throughout their lives because they had absorbed more lead when they were young. The costs to the nation of all this lost brain power had never been estimated. They still haven't.

This study was a direct frontal assault on a claim that the lead industry, and Ethyl Corporation in particular, had relied on for more than half a century: that airborne lead compounds, at the levels most Americans were exposed to, had no demonstrated health effects whatsoever. Needleman's article pointed to a health effect that was serious, widespread and, worst of all, directly linked to gasoline. The industry's counterattack was swift and massive.

"If you ever want to be intensively peer reviewed," Needleman later remarked, "just produce a study with billions of dollars of implications and you will be reviewed to death." An industry trade group, the International Lead Zinc Research Organization (ILZRO), hired dozens of scientists, including the psychologist Claire Ernhart of Case Western Reserve University, in an effort to discredit this work. In 1982, the ILZRO charged that Needleman had committed misconduct in carrying out this study, because he had allegedly not corrected for poor mothering, social class, and measurement errors in assessing intelligence and behavior. In response to these charges, EPA formed a committee to review Needleman's study. One member of the committee was Sandra Scarr, a psychology professor at the University of Virginia. With Ernhart, she later wrote a critique of Needleman's original study, claiming that his findings had failed to take into account the effects of poor housekeeping and parenting. In their view, lead poisoning was just another instance of a social problem whose solution was for mothers and other caregivers to take better care of their children. Needleman contended that better housekeeping would not be required if the stream of exposure to lead were turned off in the first place.[5]

This would seem on the surface to be a purely professional disagreement about the interpretation of the evidence, but in 1991, Ernhart and Scarr formally accused Needleman of scientific misconduct in a letter sent to the Office of Scientific Integrity of the National Institutes of Health, and National Institute of Mental Health. Echoing the claims of ILZRO a decade earlier, they charged that Needleman had been biased in the conduct of his research, because he had not controlled for the role of family environment, social status, or other potential confounders, and had distorted his findings. At the time, their charges appeared to be leveled independently. It was not known then that their work was supported by the lead industries.

Though unfounded, these grave accusations triggered a series of costly investigations that went on for nearly a decade. In one of these inquiries, a committee working for the EPA in 1983 audited all of Needleman's files on each of the children. When this group found minor statistical errors in some of the work, at the behest of ILZRO, Hill and Knowlton, one of the world's largest public relations firms, distributed a letter to journalists saying that Needleman's work had been discredited.

In pressing her case against Needleman, Ernhart was represented by Hunton and Williams, a Washington, D.C., law firm that had previously represented lead companies. When the investigation finally ended in 1994, Ernhart told the *Chronicle of Higher Education* that her legal fees had been paid through a trust fund, but did not elaborate, saying, "I was asked to keep this matter in confidence." She insisted that accepting funds and legal counsel from the lead industries did not mean that her work was biased.[6]

In an editorial looking back on these attacks, written in 1992 for the medical journal *Pediatrics*, Needleman described the devastating impact of the searing review he underwent.

"If my case illuminates anything, it shows that the federal investigative process can be rather easily exploited by commercial interests to cloud the consensus about a toxicant's dangers, can slow the regulatory pace, can damage an investigator's credibility, and can keep him tied up almost to the exclusion of any scientific output for long stretches of time, while defending himself."[7]

Needleman spent more than ten years and thousands of dollars facing repeated challenges, including demands for all his original data, legal subpoenas to reproduce his files, and efforts to discredit him personally. Until Needleman's pioneering work, the government had mounted few efforts to study the effects of lower levels of lead on children's health, relying, as did many agencies at the time, on studies provided by industry. His study was the first of many others to show that low levels of lead over the years could have insidiously toxic effects on the brains of children, including damaged hearing, stunted growth, and impeded learning, no matter what their social class.[8] He later published a follow-up study of the children as they entered their middle-school years, showing that the effects of lead persisted into adolescence and included higher rates of criminal and delinquent behavior.[9] The fact that most of these children were Caucasian made it clear that lead was an equal-opportunity pollutant, the effects of which were not limited to poor, African American children. For his refusal to buckle to these pressures and for being "the unsung hero behind one of the greatest environmental health gains of modern times," Needleman would receive the prestigious Heinz Foundation Award in 1996. But for the better part of two decades, he endured a living hell at great personal and financial cost.

Long before Ernhart and Scarr filed their complaint with the National Institutes of Health, it was clear that much more was at stake than an effort to discredit a single scientific study or its author. A number of studies at that point indicated that chronic, low doses of lead proved toxic to the brains of children. But Needleman's work stood out because it showed that lead poisoning was not solely restricted to the poor.

In 1980, Ronald Reagan had been elected president on a promise of sweeping reforms of government, and EPA ranked pretty high among his immediate targets. Within a week of his inauguration in 1981, President Reagan reversed thousands of the Carter administration's recent personnel appointments, including a number of new hires at EPA.

The president's new crew at EPA was headed by Administrator Ann Gorsuch. She eagerly took on the mandate to curb federal excesses in regulation. One of the first rules that Vice President George Bush's Task Force on Regulatory Relief targeted for scrutiny was the proposed phase-down of lead as an additive in gasoline. This regulation had been put on hold during the oil embargo of 1979–1980, but by 1981 oil was plentiful again. When the proposal to remove lead from gasoline resurfaced, scientists working with industry, including Ernhart, expressed doubts about its merits. As EPA staffers prepared to issue this regulation, Ethyl made one last, bold foray. The company sent a team to meet privately with Gorsuch to explain how requiring a drop in lead would bankrupt refineries, harm engines, and not really protect public health. According to those who attended this meeting, Gorsuch listened attentively and gave a somewhat unusual signal of her intentions. When asked whether she had plans to enforce the proposed regulations to phase down lead, she reportedly winked.

More than a thousand EPA employees, including many in enforcement, lost their jobs in the first months of the Reagan administration. Within the year, so did Gorsuch. She resigned amid a scandal over Rita Lavelle, the political appointee in charge of hazardous wastes, who went to jail for lying to Congress.[10] But surprisingly, the lead regulations did not die. A strange brew of political conservatives and public health advocates managed to hand the Ethyl Corporation its first major setback since 1924.

The only Reagan appointee to EPA to survive the early days of the administration was a Utah lawyer named Joseph C. Cannon. Cannon set about becoming well informed on the issue, creating a small college of advisers inside and outside government whom he regularly tapped for independent opinions. I was one of them. Whenever industry scientists advised him of the safety of lead, he sought the views of his own experts.

As the father of a retarded child, the conservative political columnist George Will had a special interest in anything that could explain why children's brains do not all develop in a healthy manner. We met in 1982 at one of those Washington parties where people have perfected the art of conversing animatedly with you while they discreetly look past your shoulder to make sure they are not missing a chance to collar somebody important. We began to argue right away. He staunchly defended efforts to get the government out of private business. Still, he was fascinated to hear about Needleman's work and the long tradition of literary portrayals of lead's damaging properties. I sent Will all the documents I was then reviewing as a member of the Centers for Disease Control Advisory Group on Lead Poisoning Prevention.

As one of Washington's most widely read and well-connected writers, Will had opportunities available to few others. He had lunch with President Reagan and told him privately what he would later write in his *Washington Post* column: Lead had to be removed from gasoline. The very term *leaden* signified a dulled intellect. The studies Will had seen made it clear that levels of lead that did not cause any obvious harm to a child could permanently reduce intelligence and probably also worsened many other subtle problems, both emotional and physical. This proved a bombshell. If someone like George Will was weighing in on behalf of getting lead out of gasoline, even the skeptics in Congress had to listen.

The final straw was provided in 1983 by two elegant, simple, and powerful studies. Jim Annest, Jim Pirkle, and their co-workers had shown that the amount of lead in the blood of children from 1973 to 1980 had dropped dramatically in a sample randomly drawn for the U.S. Centers for Disease Control in Atlanta.[11] EPA's Joel Schwartz, then ten years away from receiving a "genius" grant from the MacArthur Foundation for his work on environmental health, and his colleague Hugh Pitcher also produced a simple graph correlating the levels of lead found in children's blood throughout the country with those used in gasoline the year before. Between 1976 and 1980, lead added to gasoline dropped 50 percent. Over the same time, lead in blood fell 37 percent. A more sophisticated statistical analysis made clear that this was not merely a coincidental correlation. Race, social class, and other measures of family status made no difference. The greater the amount of leaded gasoline sold in an area, the higher the amount of lead in children's blood.[12] A crude calculation of the savings on medical care and car maintenance tied with reducing lead ran to more than $700 million. Between 1975 and 1984, lead in gasoline fell 73 percent, and lead in air dropped 71 percent.[13] If you wanted to get lead out of children's brains, the solution was simple. Keep it out of gasoline.[14]

Neither Pirkle nor Schwartz knew that John Goldsmith, of California's Health Department, had predicted this relationship back in 1967. His article with Alfred C. Hexter included a table showing a close tie between airborne and blood lead. The two scientists warned that Los Angeles's levels of lead in the air were about twice those of other metropolitan regions, and that people faced serious risks from this exposure to a known toxic heavy metal.[15]

In this work, Goldsmith refuted claims made by Robert Kehoe, a senior scientist for the Ethyl Corporation, on behalf of the lead industry that canned food would be a much greater source of lead than the air. Kehoe had conducted experiments in 1964 on four persons exposed to airborne levels of 10 and 150 microns of lead for between 10 and 73.5 hours per week for a month.[16] We do not know who the experimental subjects were or where they came from, but by the end of the experiment their blood contained levels of lead now understood to be toxic.

Taking the information from this industry study, along with reports from a federal government effort to monitor air pollution, Goldsmith and Hexter produced a graph stunningly similar to the one Schwartz and Pitcher created some fifteen years later. Their graph sent a thunderbolt through the EPA.[17]

The rest is history. In short order, Joe Cannon and his band of policy analysts showed that conservative politics could make for sound environmental policy, if the playing field were made more level. A simple calculation of the costs and benefits of keeping lead in gasoline compared to those of removing it showed that the country would save billions of dollars and that parents and children would be saved incalculable heartache.[18] By 1985, the lead industry finally lost the battle to promote the use of lead in gasoline as safe, efficient, and essential. Following the actions of Japan and Sweden in the 1970s, the United States and most other industrial countries proceeded a decade later to phase down the addition of lead to gasoline. By 1994 less than 2 percent of all gasoline in the United States contained lead, though all that sold in the

Middle East and in much of Africa and Asia still included this deadly neurotoxin.

The struggle to get tetraethyl lead out of gasoline had taken more than sixty years. Herbert Needleman's work was brutally assaulted, subjected to an industry-financed disinformation campaign just as sophisticated and merciless as the one that kept the tobacco firms out of the reach of regulatory agencies for years.

I learned just how insidious the process of peer review[19] could be as a member of various scientific advisory panels during the Reagan and Bush administrations and as an official of the Clinton administration in the Department of Health and Human Services. From 1982 to 1985, I served on the National Toxicology Program (NTP) Board of Scientific Counselors—the group that regularly scrutinized the detailed laboratory studies of chemicals tested in animals. Our deliberations were methodical and plodding, as we examined reports of two-year-long studies of animals exposed to different doses of toxic chemicals under controlled conditions. The reason for the program was not merely to improve the science of toxicology but to provide the public with some reliable indication of the potential of various agents to affect their health.

The underlying premise of toxicology studies is that what happens to animals under controlled conditions predicts human responses in the real world. On this assumption, the World Health Organization's International Agency for Research on Cancer (IARC) created a system for reviewing information on the potential of chemicals to cause cancer, classifying them as either confirmed or suspected carcinogens, or as agents that cannot be categorized reliably. In the United States, the so-called *Annual Report on Carcinogens* was begun in 1978 to produce a list of agents believed to induce cancer in humans. (There were 218 of these agents in 2000.) The list is broken down into two categories of chemicals, those reasonably anticipated to be human carcinogens because they are known to cause cancer in animals, and those for which we do not have enough information to say one way or the other whether they produce cancer. The annual report is a sort of reverse

Academy Awards system for chemicals. Companies fight furiously to keep their products off the list. So far, ten editions of this "annual" report have come out in twenty years—testimony to the protracted process underlying the listing and ranking of materials.

On its face, the process we went through at the NTP was pretty simple. Scientists who had conducted laboratory studies would present charts, tables, graphs, slides, and various statistical measures from laboratory studies of rats, mice, and animal and human cells, all showing the biological consequences of exposures to specific agents. Of course, the unresolved issue remains the extent to which studies conducted with animals and cell cultures predict what will happen to humans with similar exposures. At the NTP review sessions, before deciding to list a chemical as carcinogenic, we looked at damage in specific organs, metabolites of the chemicals being studied as they moved through the body of animals, and the health of all their tissues. The decision was never based just on statistical measures. But how did the agents being tested get picked for study? Who decided what tests would be conducted? We never heard about that.

I first realized there were a number of agendas at play when we were slated to review a chemical called trichloroethylene (TCE) in 1983. All the earlier studies on this compound had been found to be flawed. Some of them did not use enough animals to come up with statistically significant findings. Others had not carried out acceptable toxicology studies for long enough periods to approximate the lifetime exposures of humans. Others had not been completed, because the animals died in the middle of the study. Somehow or other, after more than thirty years of TCE use and at least a dozen different studies, scientists still could not say whether TCE caused cancer or other problems in workers.

American manufacturers and other businesses were using 100 million gallons of the stuff a year—about two quarts for every person in the country. At room temperature it is a nonflammable, colorless liquid, with a somewhat sweet odor and a sweet, burning taste. Mainly used to remove grease from metal parts, TCE was also part of such household products as typewriter correction fluid, paint removers, adhesives, and spot removers. It also happened to be the most widely found contaminant in federally listed hazardous waste sites.

The NTP's meetings are open to the public. As our deliberations on the huge files of information on TCE were about to begin, the room suddenly flooded with folks who did not look at all like scientists. A group of men entered, dressed in well-tailored, dark suits and carrying very prosperous-looking briefcases. They took notes on everything that was said and saw to it that within two weeks of our meeting, memos written by other experts addressed all of the issues we raised. The toxicological studies and the human evidence before us were all, according to these experts, so flawed as to be useless. Based on the evidence before us at that time, the board could not reach a conclusion about whether TCE caused cancer in humans.

The controversy over TCE continued for two decades. By 2000, there were more than eighty different studies in humans and several hundred in various animals, including living tissues and intact organisms. Of course, not all of the studies in humans showed evidence of harm and not all of them were of good quality.

We will never know how many studies that could have been carried out were never undertaken, but we know of at least one instance in which such research was deliberately made impossible. Two families that once resided opposite a Xerox manufacturing plant in Rochester, New York, received $4.75 million after filing suit charging that TCE had contaminated their groundwater and soil, causing cancer in a five-year-old girl. As part of the resolution of this case, they left their homes and closed their health records. Information on these cases remains secret and sealed.[20]

Based on the sum of the published work at hand, including an unusually well-done study on German workers with detailed information on exposure, the scientific staff of the NTP had unanimously recommended that the formal evaluation be changed: TCE should be listed officially as a human cancer-causing agent on the listing of confirmed human carcinogens. The NTP held a review meeting to consider this recommendation in December 2001.

Here is what the Halogenated Solvents Industry Alliance (HSIA), the group that represents Dow Chemical and other major producers of TCE, boasted in its Web posting in January 2002.

A bold headline stated, "Panel Rejects Upgrade for TRI."

The text trumpets a victory on behalf of TCE:

> A group of science advisers to the National Toxicology Program (NTP) voted 9 to 1 in mid-December to reject a nomination to upgrade the carcinogenicity classification of trichloroethylene (TRI) to "known human carcinogen." . . . The nomination was based on the results of controversial studies of German workers that have been challenged by many in the scientific community.
>
> HSIA submitted extensive comments opposing the upgrade and arranged for several prominent scientists to speak at the December meeting. In its comments, HSIA noted that epidemiological studies of American workers exposed to TRI show no consistent evidence of a relationship with cancer. This position was in agreement with the views expressed by the independent epidemiologists and toxicologists who spoke at the meeting.[21]

To read this posting correctly you have to understand two specialized terms. *Controversial*, in this context, means "not favoring our position," and *independent* means "testifying on our behalf." In preparation for the meeting, HSIA had recruited some of the world's top epidemiologists, including Dmitri Trichopoulos, professor and Vincent Gregory Chair of Epidemiology at Harvard University, and Hans Olov Adami, chairman of the epidemiology program of the Karolinska Institute in Sweden. They did not testify in person, however, but responded in writing to HSIA's carefully phrased questions. Their written report was submitted as part of the record by HSIA's representative. HSIA asked these two distinguished professors to restrict themselves to the narrow question Did the studies in *humans alone prove* that TCE caused cancer?

The professors rejected the opinion of NTP scientists and an analysis by Daniel Wartenberg, an epidemiologist at Rutgers University. Wartenberg had combined more than seventy studies that looked at people exposed to TCE and that he judged to be well conducted; he found increased risk of kidney, liver, and cervical cancer and probably others as well.[22] The professors' rejection of Wartenberg was straightforward. They would not accept any aggregate analysis of the data from multiple studies, each of which was imperfect in some way. They insisted that only faultless studies merited inclusion in any summary of the work. If they took each flawed study strictly on its merits, they would throw it out. The remaining studies would only be those that showed no effect of TCE. Among the studies that the professors rejected were two that had found that workers exposed to TCE had between eight and nine times the risk of kidney cancer, compared with those without such exposure.

The professors offered two major reasons for dismissing this overall work. First, most of the discarded studies lacked good information on exposure to TCE alone, since workers were typically exposed to many different solvents. Second, many of the studies showing an effect involved small groups of workers. One investigation that Trichopoulos and Adami dismissed had found that German workers regularly exposed to such high amounts of TCE that they nearly passed out from the fumes had nearly ten times more kidney cancer than those without such exposures. The professors called this study an outlier because the risk of developing kidney cancer in these workers was so high. But if other studies were found lacking because they did not have good information on precisely what workers were exposed to, why throw out the one study where this information on exposure was unequivocal—and provided clear evidence of human harm? As Elihu Richter, an occupational medicine specialist from Israel, has remarked, "Hey, buddy, that outlier you just threw out of your study happens to be my patient."

Another study rejected by Trichopoulos and Adami had examined fewer than two hundred workers—again with relatively high exposures—who developed eight times more cancer than those without such exposure. In this case, there was too little information on what other chemicals the workers had been exposed to.

In taking this stand, the professors are subjecting environmental studies to the same standards that apply to controlled studies of new drugs. If you are supposed to be testing a new painkiller and during the trial you also take lots of aspirin, whatever analgesic effect may occur in you doesn't tell you anything about the new painkiller. If everybody does as you did, the whole study is useless. The professors are precisely correct that epidemiologic investigations of workplace hazards seldom allow us to conclude which of the thousands of chemicals in wide use accounts for the detected excesses in illness or death that can be found in some factories. But the absence of definitive evidence on specific exposures should not be confused with proof that no such harm exists. The real world in which epidemiology is conducted is messy and hard to control. Exposures are seldom well characterized, nor do they usually occur from a single compound.

Wartenberg disagrees with Trichopoulos and Adami concerning evidence on the dangers of TCE for humans. He believes, as do I and numerous researchers inside and outside the federal government, that studies in cell cultures, rodents, and other experimental models of TCE should be included in any assessment of the compound. We know that TCE can damage genes, impede brain function, diminish reaction time, and induce liver and other cancers in laboratory animals. Wartenberg has combined all the studies ever done on TCE, ranked them as to quality and type of analysis, and concluded that they do justify classifying TCE as a cause of cancer in humans. How can there be such a broad disagreement within a single field?

Adami and Trichopoulos's work for the solvents industry is at the center of a raging debate in epidemiology today. They have confined themselves to the narrow question, narrowly considered, of whether we have a particular kind of proof of human harm. They pointedly ignore the extensive experimental literature showing that TCE damages genes and cells and induces cancer in animals. Their view is that the best study designs in epidemiology today involve case-control studies, in which individuals with a given disease are compared in great detail to a carefully matched set of people without the disease, or where those who have been exposed to a given chemical are matched for comparison with those who have not been so exposed. The solvents industry, with the help of Trichopoulos and Adami, has taken the position that because evidence on humans gathered in the best possible manner is inherently unavailable for TCE, all other types of information, no matter how persuasive, must be ignored. On this question, the perfect must forever remain the enemy of the good.

In December 2001, the NTP agreed with them.

Another instance of this contentious and protracted review process occurred with butadiene. Critical to the production of synthetic rubber for the U.S. war effort, butadiene plants were financed by the government in the 1940s to make up the shortfall created when the Japanese cut off exports of rubber from Southeast Asia. Yearly production jumped from 3,300 metric tons in 1941 to more than 830,000 metric tons by 1945 and was nearly 2 billion pounds at the time of the NTP assessment of the compound in 1983.[23] When the NTP review was being conducted, workers could legally be exposed to 1,000 parts per million (ppm) of butadiene in the air they breathed, essentially indefinitely, according to guidelines established by the American Council of Government and Industrial Hygienists, a voluntary group of industrial and governmental experts. By 1994, butadiene would be among the top twenty most commonly produced toxic chemicals in the United States, with more than 3 billion pounds. Not until 2001, after years of expensive studies that remained inconclusive by design, was butadiene officially declared a human carcinogen.

Back in 1983, the 1,000-ppm exposure limit set for butadiene was intended to prevent irritation to the eyes and upper respiratory tract—it was not aimed at cancer. By 1985, other NTP studies showed that mice exposed to these same levels for two years developed tumors throughout their bodies. This was a bombshell. Usually, the exposures tested in animals over two years are much higher than anything humans would typically encounter on a regular basis. Researchers choose these exposures to give the animals doses that approximate those that people can receive in their seventy years of life. In the 1985 study and others, however, we had evidence that a widely used industrial agent induced tumors in animals after two years of exposure to levels that could legally be found in the workplace.[24]

In 1987, I published a report in *Lancet* on these findings, urging that companies immediately lower the levels of butadiene to which workers were being subjected.[25] In response to the evidence of increased cancers in animals, the International Institute of Synthetic Rubber Producers questioned the appropriateness of using studies in rats and mice to estimate human impacts. Yes, they said, the studies by the NTP at levels of 1,000 ppm and higher had found increased rates of cancer in many different organs in rats, including the pancreas, uterus, Zymbal gland, mammary gland, thyroid, and testis, and, yes, they had also found cancers in other organs in mice, including lymphomas and cancer of the heart, forestomach, lung, liver, mammary gland, ovary, and preputial gland. The industry-supported scientists pointed out that humans do not have some of these organs and have different enzymes and proteins. They also said the doses used in these studies were so high that these alone might have created unusually broad numbers of cancer. In the meeting in which the limits were recommended, Ted Torkelson, then at Dow, argued that the high doses alone had induced these cancers. We know now and knew then that high doses of anything can kill you. But they do not give you cancer.

By 1985, the NTP completed another study on butadiene using much lower doses. Different groups of rats and mice had continuously breathed 6.25, 20, 62.5, 200, and 625 ppm for up to two years. All animals exposed to any level of butadiene got cancer. The lowest levels caused rare heart and lung tumors. Independently, one study lasting only thirteen weeks found tumors in many different organs, whereas another showed that, once inhaled, butadiene converted into some nasty-looking compounds, which made it a very active agent. In 1996, the Occupational Safety and Health Administration officially lowered the permissible exposure limit a thousandfold, from 1,000 ppm averaged over eight hours to 1 ppm. Not until its *Ninth Annual Report on Carcinogens* in 2001—when global production reached more than 18 billion pounds—did the NTP finally label butadiene a known human carcinogen.[26]

This listing came about despite the rubber industry's attempts to stifle research on this question. In 1986, after several years of working on his doctoral dissertation at Johns Hopkins University under Genevieve Matanoski with support from the rubber industry, Carlos Santos-Burgoa, a medical doctor, had carefully assembled a cohort of butadiene workers. He had confirmed their exposures, tracked their health status, and found that they had more than six times the risk of cancers of the blood-forming organs. The researchers were stunned when the rubber industry, in an effort to keep this information from becoming public, threatened to sue the university if Santos-Burgoa used these data in his dissertation.

The Donora Wire Mill in 1910. At this point the town was barely settled. European immigrants came to a place already made black and smoky by its mills.

The Donora Steel Mill in 1948, during the fatal inversion.

A sketch of Donora made by Charles Shinn for the U.S. Public Health Service in 1949. The crosses indicate the locations of deaths that occurred during the killer smog. The zinc plant is just inside the horseshoe bend in the Monongahela River.

Views of Donora's steel mill and wire works taken from across the river in Webster, Pennsylvania. All of these photographs are made at noon on various days in 1948. On the worst days, the town was completely obscured by smog.

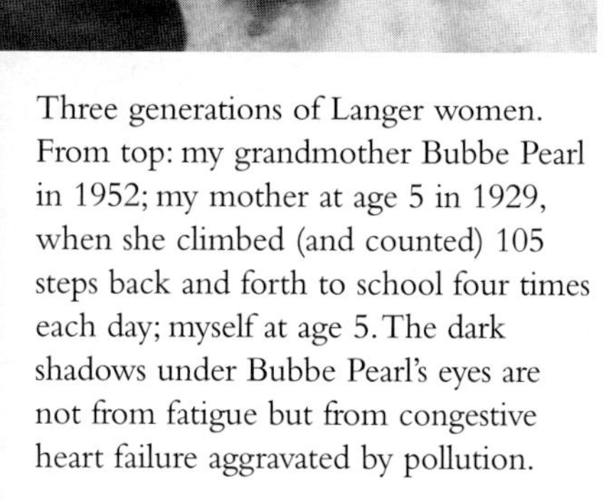

Three generations of Langer women. From top: my grandmother Bubbe Pearl in 1952; my mother at age 5 in 1929, when she climbed (and counted) 105 steps back and forth to school four times each day; myself at age 5. The dark shadows under Bubbe Pearl's eyes are not from fatigue but from congestive heart failure aggravated by pollution.

Central London during the killer smog, December 1952. At this point, visibility is less than 30 feet. During the height of the smog people could not see their own hands or feet, and buses had to be led by policemen walking with flares.

"Night At Noon." London's Piccadilly Circus at midday Sunday, January 16,1955.

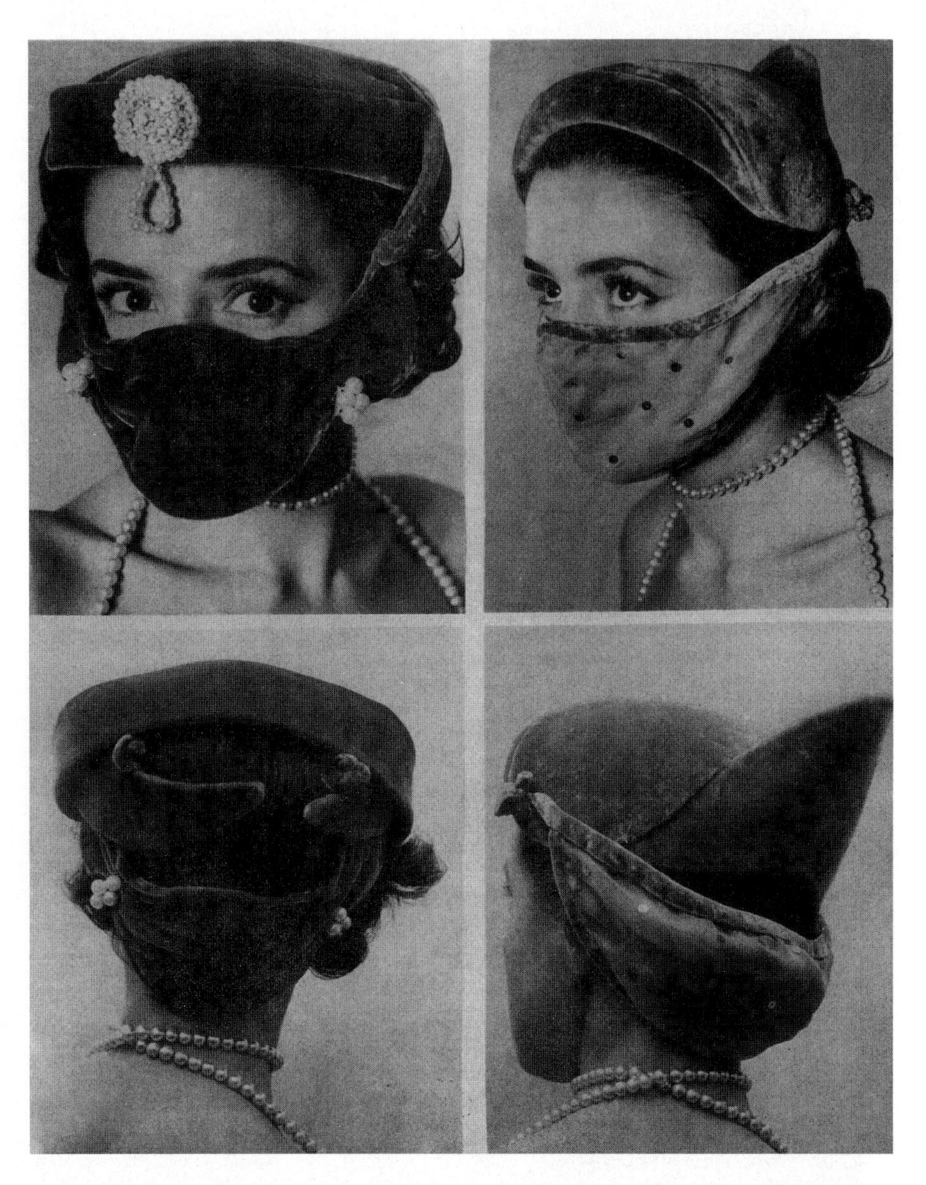

Designer Millinery and Smog Masks by Lady Newborough, 1953, proved useless against the smog's effects.

Protests at Pasadena City Hall on November 9, 1954, following fifteen days of smog in October.

Dense fog over the Los Angeles Civic Center, 1955. Note that the buildings project above the base of the inversion layer, while the smog remains below.

New York City Mayor Robert Wagner (right) shows Sen. Edmund Muskie. a smoky view of New York, as Muskie conducts a hearing on air pollution at City Hall February 18, 1964. Muskie, chairman of the Subcommittee on Water and Air Pollution, opened the hearing by saying that the New York City region poses the "worst air pollution problem" in the nation.

John Goldsmith, 1970. One of California's pioneers in environmental health research.

Eric Stork, 1980. A self-described "nameless, faceless bureaucrat" of EPA who helped bring the car companies to their knees in the 1970s and '80s.

Mary Amdur in about 1952, when she was fired from Philip Drinker's laboratory at Harvard University, for her refusal to withdraw from public presentation her findings on the hazards of acid aerosols. Courtesy of David Amdur.

Lester Lave, the impossibly young looking economist who started a revolution in environmental health research in 1970.

EPA Administrator William D. Ruckelshaus and President Richard Nixon at the signing of the Clean Air Act, December 31, 1970. (Remington's painting of Teddy Roosevelt in "The Assault of San Juan Hill" in background.)

A 1927 ad for Ethyl Gasoline that makes no mention of the word lead. In a public hearing two years earlier, Frank Howard, an executive with Standard Oil, called leaded gasoline "essential in our civilization; a gift of God."

An Ethyl Gasoline ad from 1929. The ads extol high compression and do not indicate that the fuel contained lead.

DOONESBURY 7-11 UNIVERSAL

Public health innovators, clockwise from top:
David Bates, H. Leon Bradlow, Jack Spengler, Herbert Needleman.

Mexico City photographed from the same point at three different times of the day in 1999, as smog builds up and obscures the city.

Front page picture from *Reforma* newspaper, September 21, 1999.

Clockwise from top:
Victor Borja-Aburto, Lester Breslow, Mario Molina, F. Sherwood Rowland.

Tom Mangelson reports that in 2002, pack ice in Svalbard, Norway, was hard to find. Without ice flows from which to hunt seals, polar bears are not able to obtain enough food to survive. This series shows a 1500 pound adult bear trying to leap to ice. Dens of cubs have reportedly collapsed due to melting conditions. Photos courtesy of Thomas D. Mangelsen, www.mangelsenonline.com.

The Defiant Inmate, Dachau Concentration Camp Memorial. An inscription on the statue reads, *Den Toten Zur Ehr, Den Lebenden Zur Mahnung (for the Dead an honor, for the Living, a warning),* photo by Luis Cifuentes.

The suit never materialized, but the industry did withdraw all funding for the Johns Hopkins research, transferring its support to the University of Alabama, which completely redid the work. In 1991, Santos-Burgoa got what he remembers as an unexpected birthday present at an early breakfast meeting with Philip Cole, who had led the June industry assessment by the University of Alabama team. "We have completely redone your work. There is no way to get rid of the risk that you have found. We have it in our data, too."

Santos-Burgoa recalls, "For years, we wondered whether we had done something wrong. Here we had confirmation from someone who had been brought in to demolish our work that we had been right all along. This was bittersweet for us."

It was more than bitter for the thousands of rubber and tire and plastics factory workers who labored under higher-exposure conditions while all this was being resolved. We cannot calculate precisely how many of them will get cancer because efforts to reduce exposures were delayed, but we can be sure it will happen.

Having failed in its wide-ranging efforts to keep butadiene from being labeled a human carcinogen, the industry set its sights on the international arena. The International Agency for Research on Cancer (IARC) of the World Health Organization also holds review meetings to create their own list of known and suspected carcinogens. Something without precedent happened in 1998. At the IARC Task Group meeting, where the evaluation of butadiene was made, the group voted seventeen to thirteen to classify butadiene as a human carcinogen. Then, according to private accounts, one scientist who had voted in favor of this classification left the meeting right afterward to return home.

Only a few people can say what happened that evening, and they aren't talking. All that the rest of us know is that the very next day, something extraordinary took place. Without any record of what went on or why, the vote on how to list butadiene was repeated. No such day-after revote had ever occurred on an IARC panel before, and none has since. By fifteen to fourteen, butadiene was downgraded from *confirmed* human carcinogen to *probable* human carcinogen. The panel chairman, Benedetto Terracini, asked IARC to disclose, in its report of the decision, how close the vote was. He later complained when the agency failed to do so.

Whatever happened behind the scenes to change the listing of butadiene appears to be part of what Lorenzo Tomatis, the founding director of the IARC program to evaluate carcinogens, deems a disconcerting pattern. "One may question," he writes, "whether interests other than purely scientific were possibly involved."[27] In the past, if a studied exposure caused cancer in lab animals, it was regarded as a threat to humans. No more. Yes, animals do develop liver cancer—an always fatal disease—when exposed to some of the widely used plasticizers called phthalates. But liver tumors arise by different pathways in animals and humans. As a result, the agency has just changed the listing of this high volume industrial material from "probable human carcinogen" to "not classifiable as to human carcinogenicity." Atrazine, one of the most widely used pesticides and frequently encountered contaminants in rural groundwater in this country, has also been downgraded, based on a similar logic.

A whole roster of industrial carcinogens stands poised for such relabeling. Of course, science has made impressive progress in understanding how cancer arises, and people certainly do differ from lab rats. Still, everything proven to cause cancer in humans has been shown to do so in animals. So how can we be sure that we are different enough? Tomatis concludes that:

> Ominous consequences on public health may follow if such hypotheses [that animals and humans diverge in how they get cancer], once actually tested experimentally, are shown to be incorrect, or they do not account adequately for the wide range of susceptibility . . . in human populations. The revised evaluations may therefore open the door to free exploitation of products, the production and use of which should instead be strictly regulated or banned.[28]

By the time that we are able to learn whether Tomatis's concerns are well founded, he and I and most of those reading this book will no longer be around. It all comes down to a question of what risks you are willing to accept. Is it better to err on the side of protecting public health, or on the side of promoting industrial growth? There is no free lunch.

When she first began developing new models for public health research in the 1950s, Mary Amdur pointed out the limits of epidemiologic research and the need to learn from animals. Human studies can never be used to predict what will happen, and they seldom allow us enough information to understand what has taken place with respect to diseases that can take many years to develop. She understood that epidemiology is best suited to confirming past hazards and cannot be used to predict future ones. Trichopoulos and Adami's position on TCE, for example, is perfectly consistent with a view of epidemiology as an assenting science that tries to reconstruct the past in order to understand the present. But there is another goal for public health research: predicting and thus preventing future harm. This purpose takes us out of the realm of pure knowledge and into an arena in which lives are at stake. Our standards of proof must change accordingly.

Those who spend their lives discovering links between health and the environment can easily be seduced by the complexity of the analysis. The elegant equations employed to describe patterns of health and the environment can obscure the reasons for asking these questions in the first place. The basic reason for making forecasts is not to be proven correct. Public health activities should be devised to prevent damage, not to confirm later on that harm has happened.

Saying that things are complicated should not be misinterpreted as meaning that they cannot be understood at all. We now know that many environmental health hazards, whether those of the workplace, air, or water, cannot be studied with the case-control design or be easily put into simple statistical formulations. For many common low-level pollutants, there simply is no unexposed or less-exposed control group; everybody has to breathe. Everybody has to drink water. Most of us have to work somewhere. Over the past forty years, studies in more than twenty countries have all showed similar trends. These studies have found that specific types of air pollutants, include fine particles and gases of sulfur, carbon monoxide, and nitrogen, directly worsened a whole range of health problems. How could epidemiologists be sure these findings were due to a causal connection between air quality and poor health?

Figuring out whether observed differences in public health are real or merely haphazard always involves messy, large-scale data. Statisticians start out assuming that there is no difference between any two observed groups. They use a variety of mathematical models for measuring the importance of any difference that occurs.

A rate, such as a death rate, consists of a number of events relative to a standard-sized population, usually 100,000 or sometimes 1 million. Because districts within cities, such as zip-code or voting areas, do not come in uniform sizes, creating rates can be problematic. If you are interested in the rate of some event in the overall population, you need to know how many people of all ages reside in a given area. For determining the rate at which an event occurs during the first year of life, you need to know the normal number of infants born and living to the age of twelve months. For infants, rates are often expressed in terms of the number of events per 1,000.

When we look at patterns in human populations of any age group, we rely on two basic principles to see whether what we have observed is important. First, where the differences between any two groups are large, they are less likely to be random. Second, your chances of finding a significant difference are greater the more times or things you observe. This is called the law of large numbers. The two principles are related: The larger the sample, the smaller an observed difference can be without being probably random. Roughly speaking, statistical testing asks what the chances are that the observed differences are something other than random. If it is sufficiently unlikely that the difference would have arisen by chance alone, then the difference is called *statistically significant.*

More precisely, statistical testing assumes that the two groups are identical in all important aspects and then mathematically calculates the chances that you would see a difference between groups at least as big as the one you actually saw. This probability is called the p value. P can range from 0 to 1. A value of 1 means there is a 100 percent chance that the findings are accidental; .5 indicates there is a fifty-fifty chance that the findings have happened accidentally; and .05 signifies that there is only a 5 percent chance, or one in twenty, that the findings are a random occurrence. According to scientific convention, a value of .05 is usually accepted as statistically significant. But this is merely a convention. In physics the upper limit for statistical significance is $p = .1$, meaning that there is a 10 percent chance that the results could have happened randomly. But public health demands a *p* of .05, that is, tougher statistical evidence.

Where small numbers of persons or very rare events are involved, using the *p* value may not be appropriate. In these situations, epidemiologists rely on confidence intervals to show the range of values that could have occurred. Basically a confidence interval is that range within which a given result is likely to fall about 95 percent of the time.

Statisticians consider differences important if they achieve statistical significance. This issue plagued the early work on air pollution. Of course there were differences in deaths and illnesses during times of higher pollution, but how could researchers know that these differences are not just due to life's normal variations? The answer is that any given set of phenomena can be expected to follow a pattern. Despite the uniqueness of each illness and death, these events have some things in common, and statisticians work on those common characteristics. Did all the people who died or became ill live in a certain area? Did they sicken on a certain date? Do these cases occur at about the same time of day or month? Are there times when they do not occur at all? Is there something that each of these single sad events has in common with the others and with the physical region in which they occur? If conditions have changed in a given region where the rates were high, does this change appear to affect the rates in any way?

With the answers to these questions, epidemiologists can create an array that may show whether these events form a particular pattern in time or space. Looking at similar occurrences together permits the epidemiologist to notice whether some common circumstances could characterize each of these losses. It is the test for statistical significance, operating on these observed patterns, that changes them from hypotheses and intuitions to something approaching fact.

Another thing public health scientists look for when deciding whether they have found a true connection between the environment and health is any evidence that with more exposure, you get a stronger effect. This is referred to as a relationship between dose (or the amount of exposure) and response (or the health endpoint under study). For instance, we know that those who smoke two packs of cigarettes a day tend to get sicker sooner and die younger than those who smoke one pack a day. But the real world throws us lots of curve balls. For instance, those who smoked four packs a day in the 1950s actually lived longer than those who smoked a bit less. The reasons are pretty clear. Anyone with lungs and heart capable of sustaining the continual bombardment of so much smoking probably had some resistance to these toxins. Even smoking, the nastiest and most important environmental hazard we know of, doesn't kill everyone.

Bradford Hill, one of the pioneers of public health research, warned that statistical reasoning should not become the sole litmus test for public health analyses:

> I wonder whether the pendulum has not swung too far—not only with the attentive pupils but even with the statisticians themselves. . . . There are innumerable situations in which [statistical tests] are totally unnecessary—because the difference is grotesquely obvious, because it is negligible, or because, whether it be formally significant or not, it is too small to be of any practical importance. What is worse, the glitter of the t table diverts attention from the inadequacies of the fare. Only a tithe, and an unknown tithe, of the factory personnel volunteer for some procedure or interview, 20% of patients treated in some particular way are lost to sight, 30% of a randomly-drawn sample are never contacted. The sample may, indeed, be akin to that of the man who, according to Swift, had a mind to sell his house and carried a piece of brick in his pocket, which he showed as a pattern to encourage purchasers. The writer, the editor and the reader are unmoved. The magic formulae are there.
>
> Of course I exaggerate. Yet too often I suspect we waste a deal of time, we grasp the shadow and lose the substance, we weaken our capacity to interpret data and to take reasonable decisions whatever the value of [*p*]. And far too often we deduce "no difference" from "no significant difference". Like fire, the [chi-square (a test of statistical reliability)] test is an excellent servant and a bad master.[29]

This sage analyst warns against an overly narrow view of epidemiology. He notes that real life seldom permits the luxury of waiting for all the information to become available before decisions have to be made:

> In occupational medicine our object is usually to take action. If this be operative cause and that be deleterious effect, then we shall wish to intervene to abolish or reduce death or disease. While that is commendable ambition it almost inevitably leads us to introduce differential standards. . . . Thus on relatively slight evidence we might decide to restrict the use of a drug for early-morning sickness in pregnant women. If we are wrong in deducing causation from association no great harm will be done. The good lady and the pharmaceutical industry will doubtless survive.
>
> On fair evidence we might take action on an apparent occupational hazard. For example, we might change from a probably carcinogenic oil to a non-carcinogenic oil in a limited environment, and without too much injustice if we are wrong. But we should need very strong evidence before we made people burn a fuel in their homes that they do not like or stop smoking the cigarettes and eating the fats and sugar that they do like. In asking for very strong evidence, I would, however, repeat emphatically that this does not imply crossing every "t," and swords with every critic, before we act.
>
> All scientific work is incomplete—whether it be observational or experimental. All scientific work is liable to be upset or modified by advancing knowledge. That does not confer upon us a freedom to ignore the knowledge we already have, or to postpone the action that it appears to demand at a given time.
>
> Who knows, asked Robert Browning, but the world may end tonight? True, but on available evidence most of us make ready to commute on the 8:30 next day.[30]

Hill closes his lecture on this topic with this advice: "In short, the association we observe may be new to science or medicine and we must not dismiss it too lightheartedly as just too odd. As Sherlock Holmes advised Dr. Watson, 'when you have eliminated the impossible, whatever remains, however improbable, must be the truth.'"

If statistical issues were the only challenges to conducting epidemiology, they would be daunting enough. The real scientific difficulties of the field have been complicated by a stream of disinformation fueled by the short-term economic interests of those who stand to profit from keeping matters unresolved. The effort to establish the science of environmental epidemiology has been plagued by a sophisticated and completely legal disinformation campaign, the full extent of which is not appreciated even by those who have been its chief victims.

One strategy is to make sure the data don't exist. Air pollution is not the only matter on which big money and big pressures have been unleashed to discourage or discredit research. For years some people have thought that the serious problems of pollution are long gone in the United States. Although most of us are enjoying cleaner environments, those who are poor and outside the mainstream do not always share in those improvements. In November 2001, Allen Silverstone, a professor of immunology and microbiology at the State University of New York's Upstate Medical University at Syracuse, was called in with an unusual request. Lawyers for more than three thousand people who claimed to have been exposed to PCBs and other toxic chemicals asked Silverstone to study the health of a small group of them, who did not have any diseases known to be linked to such exposures. Silverstone conducted several tests to see whether their immune systems had been affected by where they lived, worked, and grew their food.

The immune system is one of the most complicated parts of the living organism. It needs to be activated appropriately to get rid of infection, but if it becomes overactive, it can turn on itself. This is called "autoimmunity," and Silverstone is one of the world's experts on it. He agreed to look into the samples.

The site of the lawsuit was Anniston, Alabama, an area with high, unexplained cancer rates that sits in southwest Calhoun County. It is the location of a PCB plant, owned and operated since 1933 by Monsanto and recently sold to Solutia. Silverstone had never heard of the place. But that would soon change. From the 1930s to 1971, PCBs had been manufactured in this small southern county. Production involved pumping chlorine gas under pressure into a crude mixture of biphenyls and then boiling off the resulting PCBs. The residue was called still-bottom, a thick, oily, turgid muck that settles to the floor of the metal containers in which it is made. Over a hundred million pounds of this material, rich in chlorinated organic matter, was hauled south of the plant and dumped into an unlined open pit that sat right over bedrock.

For more than fifty years, the mix had been moving into local streams and into the ground.

The samples were drawn under controlled conditions and sent to the laboratory, with no personal information provided to those running the tests. Silverstone was shocked at the results: At least half of the supposedly healthy people showed altered ratios of immune cells consistent with greater susceptibility to cancer, infection, and autoimmune diseases. And these were the survivors. "The thing that blew me away was the fact that the only thing that correlated with these abnormalities was the higher levels of PCBs measured in their blood four years earlier. Not smoking, not sex, not race. Nothing else could explain what we found," Silverstone told me.

Nearly all of the people selected for this study were, in fact, not healthy—at least by the criteria Silverstone and his colleagues apply when looking at immune systems. Testifying for the defense of Monsanto/Solutia in this case was Stuart Schlossman, one of the most distinguished researchers in the field of immunology. A professor at Harvard's Dana Farber Cancer Institute, Schlossman owns patents on the very tests Silverstone was using to study the Anniston plaintiffs. In the early 1980s, he was part of a team of investigators who reported on a cluster of cases of immunodeficiency in young men, designating them as some sort of "laboratory disorder," a defect in the test systems.[31] Five months later, others figured out that these cases were not the result of a testing defect but the beginning of what became known as the epidemic of HIV-AIDS.[32]

Serving in the Anniston case as an expert for the defendants, the man who missed the call on the AIDS epidemic has testified that the fluctuations found in the immune systems of the Anniston residents do not signal harm.

He has a point. Given the finding of what looks like a biological abnormality, the next question public health researchers should logically ask is, "What does this mean for people's long-term health?" This is precisely what they have been prevented from finding out.

In March of 2002, a jury in Gadsden, Alabama—a town twenty miles from Anniston—found Monsanto and its corporate successors guilty of "negligence, wantonness, suppression of the truth, nuisance, trespass and outrage." Under the state's laws, to be convicted of outrage requires "conduct so outrageous in character and extreme in degree as to go beyond all possible bounds of decency so as to be regarded as atrocious and utterly intolerable in civilized society."[33] Two of the lawsuits on behalf of the citizens of the area were settled for more than $50 million, after a series of last-minute delays. Among other things, the moneys will be used to set up an exclusion zone around the still operating plant, moving scores of folks out of their now worthless family homes. The federal government may soon mandate studies in the area, but the horses have long ago left the barns. As of this writing, litigation was still under way. The lead attorney for these cases, former U.S. Senator Donald Stewart, reports that since the lawsuit was filed two years ago, more than 120 of the 3,500 original claimants have died—about twice what would be expected.

It turns out that Anniston is not unique. Contamination has also brought other neighborhoods to the point of their own extinction. By the late 1980s, a number of poor, black towns in Louisiana, including Reveilletown and Morristown[34] in Iberville Parish, had become so badly polluted that it proved cheaper to wipe them completely off the map—move the small number of residents and shut down or tear down all the buildings—than to clean them up. That is certainly one way to be sure that no studies are mounted.

When Lester Lave and Eugene Seskin, in 1970, first produced their findings that the dirtier the air the higher the risk of deaths and illnesses, their results were questioned because they were not members of the public health community, because they relied on statistical analyses not widely used then by epidemiologists, and because they had used large-scale data with no biological measurements on specific individuals. In 1974 Harvard University began a series of studies to produce such information under the leadership of Ben Ferris and Frank Speizer. They sought to examine the impact of fossil fuel use on clinical measures of lung health in eight thousand people in six U.S. cities. After following these people and their environmental conditions for a decade and a half, Ferris and Speizer found that whenever the concentration of particulates 10 microns in diameter increased by 10 micrograms per cubic centimeter in the preceding few days, there was a 1 percent increase in the daily death rate, and associated increases in hospital admissions, emergency room visits, and restricted activities. This work appeared in the *New England Journal of Medicine* in 1993.[35]

Arden Pope, an economist at Brigham Young University with an interdisciplinary grasp of public health, then expanded on the Harvard work. Using the American Cancer Society's continuing survey of more than half a million volunteers in 154 cities for eight years, he published analyses in 1995 showing that persons living in more polluted areas had a higher risk of dying in any given year than those living in cleaner areas. In 1980 and again in 2000, John Spengler, then the director of the Environmental Science and Engineering Program at the Harvard School of Public Health, affirmed the opinion that about 4 percent of all deaths in the United States are due to air pollution each year—about 60,000 deaths.[36]

By 1997, the world of epidemiologic research on air pollution was much different from what it was when Lave and Seskin first came on the scene. Joel Schwartz, Douglas Dockery, and teams of national and international researchers had completed numerous long-term studies that included detailed medical information on individuals along with some personally monitored information on their exposures to air pollutants. Based on this work, EPA decided that year to extend its regulations to control particles of 2.5 microns or less in diameter. Just as soon as EPA signaled its intent, the American Petroleum Institute (API), the electric utility industry, the diesel trucking industry, and other industry groups mounted major assaults. They charged EPA with relying on junk science. "Junk science" is a curious phrase that has origins well outside of science itself. It was originally coined by lawyers trying to discredit ideas with which they did not agree.

Some scientists working as consultants to the API blasted the studies on which EPA based its proposals, charging that the apparent connection between pollution and health problems might be due to other pollutants than particles, or a less healthy lifestyle in some cities compared with others. Industry sued to stop the newly proposed regulations to reduce exposures to particles, challenging the legal authority (i.e., the

Clean Air Act) and the science upon which the proposed regulations rested. The Republican-controlled Senate held hearings on the EPA's proposed new rules.

Here is how *The Nation*'s David Corn described the scene:

> It was thirty minutes before the start of the Senate Environment and Public Works Committee hearing. Outside the committee room, 150 people waited in line for the thirty seats open to the public. Standing with power-suited lobbyists were a dozen young black men, wearing baggy pants and sneakers.
>
> Was this democracy at work, with citizenry of different stripes gathering to watch Congress at work? Nah. As the 9:00 a.m. starting time approached, the homeboys, one by one, handed their spots over to yet more lobbyists, who had paid $29 an hour to line-standing services to insure that they could witness Republican senators grill E.P.A. Administrator Carol Browner about corporate America's pressing concern of the moment: an E.P.A. proposal to tighten the clean air standards for soot and smog.
>
> As lobbyists found their placeholders, C. Boyden Gray, once President Bush's counsel and now the comandante of a multimillion-dollar crusade against the proposed standards, huddled with confederates at the front of the line and made sure the right lobbyists were going to make it into the hearing room. The guards began letting people in. Almost all the available public seats went to industry types. (Talk about paying for access.) Gray—imagine Ichabod Crane in a gray suit and brown shoes—started striding down the hall. "In what official capacity are you here?" a reporter asked him. "I represent a company," he said, and turned away. He did not say that he is a leader of the industry-funded Citizens for a Sound Economy or the mastermind behind the Air Quality Standards Coalition, another fine-sounding industry group battling the new rules. (Outside the building a handful of staffers from Citizens for a Sound Economy, dressed in prison stripes, were passing out bumper stickers that read, "Tell the EPA that Barbecuing is Not a Crime!") Then, as the hearing started, a staffer for the Republican-controlled committee ushered Gray into its offices. That's true access.
>
> Not since the NAFTA tussle or health care reform has corporate America assembled such a monster of a lobbying machine. The 600 firms

and trade associations in the Air Quality Standards Coalition are each supposed to pony up between $5,000 and $100,000 for the anti-standards effort, which is similar to corporate campaigns of the past (think tobacco). The anti-regulationists of the business community are waving the usual banner of cost-benefit analysis, enlisting mega-P.R. firms to produce slick ads and manufacture "grass-roots" front organizations. They are backing a phony scientific think tank and decrying the scientific basis for the standards—in a move akin to industry's larger attempt to demean as "junk science" health and safety research that is inconvenient to corporations. Once again, the question is posed: Can a public health issue be decided on its merits in Washington without the undue influence of corporate lobbyists?

At the hearing, the pull of the anti-standards crowd was apparent. Republican senators repeatedly berated Browner for crafting new thresholds that would be too costly to meet. Chairman John Chafee, reputedly the most green-friendly of Republicans, excoriated the new soot and smog standards for pushing "too far, too fast." These days Chafee is under pressure from majority leader Trent Lott, who is hoping to revive a Republican jihad against regulations; the assault on the clean air standards is but one front. Browner patiently explained over and over that the Clean Air Act compels the E.P.A. to set standards solely on the basis of public health, not the potential cost to business, and that cost considerations can later be taken into account when states try to figure out how to meet the standards. As the room emptied after her appearance, one lobbyist said to another, "Hey, they [Republican senators] asked all your questions. Congratulations." The other exec was smiling beatifically.[37]

The argument that more research is needed was invoked as an excuse for taking no action. Senator Craig Thomas, a Wyoming Republican, reiterated the need for more research, quoting the recommendation of Morton Lippmann, who had chaired an EPA advisory panel on the topic. But Lippmann himself has made clear his preference for more stringent standards than those the agency proposed.

One industry group that played a major role in fighting the proposals was the Center for Regulatory Effectiveness, led by James Tozzi. Tozzi is the same fellow who, as an OMB official in 1972, had denied requests to expand funding for EPA's air pollution research. He is the one who told EPA he was defunding the CHESS study because, whatever it was going to reveal, Congress did not need to know. Now, three decades later, Tozzi was fighting the EPA's proposals on the ground that too little research had been done.

The stakes were enormous. The EPA regulations would require all power plants and all vehicles to reduce their emissions. Through his organization, Tozzi lobbied for an amendment sponsored by Senator Richard Shelby of Alabama that would require Ferris, Speizer, and dozens of other researchers to release all their raw data: the measurements of lung function, the death certificates—in short, everything. A huge amount of information was at issue. The scientists refused to turn it over, claiming that to do so would breach the promises of confidentiality they had made to individuals to get their cooperation.

To turn up the pressure, the industry hired unemployed actors to picket Harvard's School of Public Health, garbed in white coats, holding placards saying, "Give us your data." The pressure was unrelenting. To resolve the issue, Harvard turned to the nonprofit Health Effects Institute (HEI) in Cambridge, Massachusetts, which is sponsored by automotive firms and the government. HEI assembled a panel of world-class experts in the field, some of whom had previously questioned whether particles actually caused poor health. The group redid the entire study, examined the original data looking for all possible explanations, and confirmed the basic findings. Under the leadership of Jonathan Samet, chairman of epidemiology at Johns Hopkins University School of Public Health, HEI also mounted a completely separate study of ninety U.S. cities, which found an 0.5 percent increase in nonaccidental deaths for each 10-microgram increase in particulate air pollution sized 10 microns. This effort took three years, and it is still the subject of intense scrutiny, since new statistical software issues have emerged.

I went for a brisk walk with Samet in the rain in the fall of 1998, just before he was to give an interim report on this work to Congress. He candidly admitted, "You know we need more research; that is what I said a decade ago. And ten years from now, I'm afraid I'm going to testify, 'Senator, we need more research.' That's the nature of the beast. But we really can get on with promoting more sensible policies, even while we continue to get the work done."

Ultimately, the effort to redefine the core mission of EPA failed—that time. But those in industry who are fighting tooth and nail against the regulation of air pollution still get, for now, the last word. Senator Shelby sponsored a law requiring any researchers who received federal funds to release their raw data if requested to do so. It passed in 1998. Taking this whole thing one step further, the omnibus appropriations bill that passed Congress in 2001 included a new provision that has enormous potential to politicize environmental science even further. The bill requires federal agencies to follow guidelines issued by OMB on "the quality, integrity and objectivity" of information disseminated to the public.

This seemingly wholesome initiative (who could be against equality, integrity, and objectivity?) is the brainchild of Tozzi and his Center for Regulatory Effectiveness. The poison pill is that the measure puts the OMB officially in the business of deciding what is and is not good science. Somehow as we enter the twenty-first century, the prevailing norms of science are no longer adequate. Publication in journals may reflect peer review of one sort. But science used in regulatory actions is subject to yet another form of review. It must now be pleasing to the political appointees at OMB, few of whom have much experience with, or sympathy for, real science.

With this new law, there is no longer any pretense that mere publication in a scientific journal will suffice for findings to be relied on in the development of environmental policy. The new laws place public health researchers on notice. If you publish something of direct relevance to the regulatory system, be prepared for the sort of review that can take years of your life, keep you from doing other work, and may never be fully resolved. If you can stomach that, there's still another hurdle: Even if your work is published in the best journals in the United States, is widely acclaimed as groundbreaking and important, and is replicated by others, if a government agency wants to use it to set policy, then the accountants in the OMB can still declare it junk science. And if they do, you can be sure that some industry-financed body with an Orwellian name like Center for Regulatory Effectiveness will be cheering them on.

The federal bureaucracy—the same bureaucracy that these same conservatives insist can't do anything else right—will keep us safe from junk science. It will even accept peer-reviewed publications as objective work, except where lots of money is at stake. In that case, scientists are advised to hope for the best and prepare for the worst.

For nearly a quarter century, it has been clear that air pollution in the United States kills between 60,000 and 120,000 people each year and sickens millions more. The good news is that air pollution levels have dropped dramatically—nearly fiftyfold—from the days when most urban areas could be smelled before they could be seen. But the full story of what air pollution means for our health is being written as you are reading this book. Using information gathered by the American Cancer Society in its voluntary survey of half a million people in 154 cities from 1982 to 1998, Arden Pope and George Thurston of New York University have linked details on good and bad habits, including smoking, weight, diet, and other important personal characteristics, with measurements of local air pollution. They found that the risk of dying from lung cancer from breathing soiled air was about equal to that of living with a smoker. With every 10-microgram increase in invisible air pollution of the sort produced by coal-burning power plants and diesel vehicles, they found an 8 percent increase in the death rate from lung cancer.[38] Each year, lung cancer kills more people—150,000—than does any other form of cancer. The disease affects 80.7 white men and 122.4 black men out of every 100,000.

The industry response to this latest report, predictably, has been to label it junk science. This time someone with the name of Joel Schwartz has taken to writing memos attacking these latest findings, posting his criticisms on the Web, and distributing them to Congress and the media. For more than two decades, a quite different Joel Schwartz, formerly of the EPA and now at Harvard, has been one of the most highly regarded analysts in this field. When memos bearing the name Joel Schwartz began to circulate, criticizing new work on air pollution epidemiology, they were accorded considerable attention. The problem is that the Joel Schwartz who criticized Pope and Thurston's work is a former California government employee with a background in auto emissions testing. Now he is affiliated with something called the Reason Institute, a right-wing think tank of public relations gurus. Having their very own Joel Schwartz critique work on air pollution confuses an already confusing situation.

What if instead of wasting time on public relations games and hindering legitimate science, we had devised a national plan to phase in pollution controls as efficiently as possible? Warren Winkelstein, working for the New York State Health Department in 1968, looked at twenty-one areas in and around Buffalo, New York, and found that deaths from lung disease more than doubled in the dirtiest areas, compared to the cleanest.[39] What if we had listened? In 1955, Peter Stocks and John Campbell interviewed men dying of lung cancer in North Wales and Liverpool, England, and learned that nonsmokers living in the most polluted zones died at ten times the rate of those living in the cleanest areas.[40] What if the British government had acted on this knowledge? In 1958, a study of 187,783 U.S. Army veterans by American Cancer Society researchers Cuyler Hammond and Harold Dorn found that the death rate from lung cancer among those living in cities was twice that of residents of rural areas.[41] What if this information had been made widely available to folks deciding where to live and to city officials thinking about what forms of transport and energy to permit?

How can we possibly calculate what acting on these studies might have meant? Why should we bother to try? We can answer the latter question by looking at what would happen if the health conditions of those living in the cleanest areas of the country in the 1960s applied to the entire nation. Bronchitis deaths would have been 40 percent lower each year; lung cancer deaths, 10 percent lower.

What if, at any point in the past forty years, these reports had been used as the grounds for making the decisions about public transport, light, heat, and energy? How many lives might have been saved? Nobody can be sure. But let's look at what we do know. Each year there are about 2.5 million deaths in this country. Your chances of dying are 100 percent. Although nobody can say when it will happen, your chances of dying early and of specific causes can be calculated. If the technologies readily available in 1980—nothing exotic, just scrubbers and electrostatic precipitators on power plants, better use of wasted heat and engine exhausts, improved efficiency, and the use of more efficient lights and energy—had been in place to reduce emissions from power plants and transport, starting that year, well over a million people would have been spared earlier deaths. How much expense, how many missed quarterly profit projections, how many inconvenienced or even downright angry lobbyists are a million lives worth?

6

THE NEW SISTERHOOD OF BREAST CANCER

Stop expecting Prince Charming. We must rescue ourselves. We are the ones we've been waiting for.

—Bella Abzug

The phone rang so early one Sunday morning in 1993, I thought it had to be bad news or a wrong number. A voice boomed through the receiver.

"This is Bella. My friends are dropping like flies. I'm gonna have a hearing at City Hall on breast cancer and the environment. March 20. I want you to be there." I had never met the woman, but I recognized that distinctive timbre, at once familiar and commanding. Why was Bella Abzug calling me to talk about breast cancer?

Bella did not have the disease, but like many in the mid-1990s, she was enraged by what was happening to her friends. She was scandalized by the growing number of treatment failures and missed diagnoses and by the unacceptably shabby medical care for poor women. During Bella's childhood, one in forty women developed breast cancer at some point in her life. By the time she was student body president at Hunter College and editor in chief of the *Law Review* at Columbia University, the rate was one in twenty. Now in her seventies, Bella wanted to know why breast cancer attacked one in nine women by age eighty-five. She was convinced that scientists had not done a very good job of connecting the disease with where and how women lived and worked, and she insisted that I should do something about this.

When Aldo Leopold wrote, "We all strive for safety, prosperity, comfort, long life, and dullness," he could not have been imagining Bella. She was a woman who consistently defied the status quo. And she did not hesitate to lash out at those who followed her, if she thought them lacking sufficient zeal or commitment. Given the way Bella treated her friends, I was glad not to be among her enemies.

As a member of the House of Representatives from 1970 to 1976 and a founder of the modern feminist movement, Bella had shaped the nation's history. Geraldine Ferraro said at her funeral in 1998 that Bella had kicked down doors so that the rest of us could walk through. In the 1950s, as a young, pregnant lawyer for the American Civil Liberties Union, Bella traveled to the South to defend a black man falsely accused of raping a white woman. She slept in bus stations when no motel would take her in. In the 1960s, she championed nuclear disarmament, organized Mothers' March for Peace, and walked with the rainbow coalitions of the civil rights movement, always wearing her trademark wide-brimmed hats. Her unrelenting rhetoric fueled the early antiwar movement. She was the first in Congress to call for the impeachment of President Nixon. In 1984, John Kenneth Galbraith wrote that "in a perfectly just republic, Bella Abzug would be president."

A few weeks before the hearing, as we sipped green tea in her apartment on lower Fifth Avenue, Bella recounted how she had spearheaded what became a global movement against nuclear proliferation. What got people mobilized, she said, was measuring nuclear contamination in something ordinary. "As soon as we learned that strontium 90 radiation got into milk—our own as well as cows'—we knew we had it nailed. . . . All we needed was to tell the mothers of the world that their children were drinking contaminated milk from bottles or breasts. There is nothing more ferocious than women protecting the young."

As far as Bella was now concerned, if she could bring down a president, end a war, and halt nuclear proliferation, she could certainly fix breast cancer. "It's got to be the environment. What the hell is happening with Long Island? Why do the women there have four times more breast cancer than in other areas? Let's find the 'milk' here," Bella demanded. "Let's find the thing that people can relate to, and we can fix this problem." She could not know then that she would lose a breast to cancer two years later.

During Bella's lifetime, public talk about breast cancer shifted as radically as did the role of women. She was born the same year that the Nineteenth Amendment guaranteed women the right to vote. At that time, cancer remained a secret scourge. When the author of *Silent Spring*, Rachel Carson, wrestled with breast cancer in 1962, she wrapped herself in the very silence she had depicted when birds no longer sang in a meadow ruined by pesticides. By the 1970s, that had begun to change. Women facing breast cancer no longer whispered privately about the disease or went off alone in despair. Two of my closest friends were diagnosed in 1976. Like many of my generation, they had grown up proud of their bodies and faced their diagnoses openly and without shame. But not all women share in the ability to be so frank about what breast cancer can do to a woman's body and her sense of self. Some still shower in the dark and change in private dressing rooms, unwilling to reveal their scarred chests.

The official records of these scars for hundreds of thousands of women are not always easy to find or to interpret, because in many states the lights have not always been turned on to find them. One of the longest functioning cancer surveillance systems in the United States, the Connecticut Tumor Registry, was begun in 1941. By the 1990s maps of national patterns of breast cancer revealed some intriguing regional patterns. Rates of the disease were highest in U.S. counties in the northeast, the Great Lakes states, and California, and lowest in the South. These maps revealed just how broad the sisterhood of breast cancer had become and raised a number of important questions.

We know that rates of many types of cancer vary widely in different countries. Women who move to the United States from Asia tend to develop the same risk of breast cancer as do those who were born here.[1] This cannot possibly come about because they suddenly acquire new

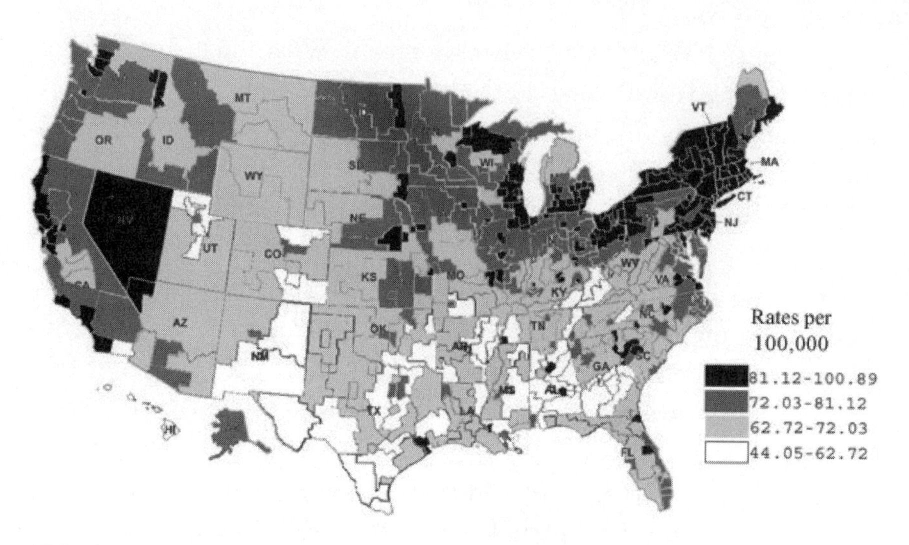

FIGURE 6.1 National breast cancer mortality map.

and more dangerous genes. What's changed is not their inheritance but their surroundings.

Why do so many more women in New York, Connecticut, New Jersey, and California develop the disease? Why do women who have lived on Long Island for forty years have four times more breast cancer than women who have lived there for only ten years? Are too many women simply eating too much, having too few children, and spending too little time breast-feeding? Was it the chemicals that decades ago had been sprayed each week on homes, schools, and gardens? How about radiation treatments for such diseases as acne and scoliosis, and the excessive use of tuberculosis screening x-rays? What about the early and prolonged use of high-dose birth control pills?

Did Long Island's and Connecticut's nuclear or other power plants, research laboratories, and industrial centers have anything to do with the patterns of the illness in these regions? What about the fact that homes in some areas of Long Island had been built on abandoned farmland that had become too contaminated to grow crops?

And who would be responsible for answering these questions?

The mere act of counting breast cancer provided controversy. First of all, people had to develop standard methods for reporting and classifying various types of tumors and their stages, ranging from early disease or stage I, through advanced metastatic disease, or stage IV. Then one had to take into account that much of the recorded increase of the disease could be the result of expanded testing for it. Remarkably, not until 1994 did the federal government issue any standards for mammographic screening, the bosom-smushing technology that looks for microscopic calcifications that can mark the early signs of breast cancer. When I visited Arkansas in 1994 as an official of the federal government, mammograms in one out of every three clinics were run by someone with little medical training, sometimes the office receptionist. The chances of getting tested for breast cancer then and now are much greater for wealthier white women than for any other groups in our country. More troubling is the fact that the newest technologies can find abnormalities in the breast, such as so-called ductal carcinoma in situ (DCIS) or small residues of calcium called microcalcifications, but cannot tell us what these mean. In the 1990s, DCIS appears to have doubled in some areas. We think that about half of these cases will not develop into cancer. But what is a woman to do until she finds out? At this point, we lack the capacity to say which tumors should be removed and which can be left alone.

There are several ways that scientists go about counting breast cancer. One way is just to ask how many new cases have occurred, relative to the number of women living in a given area. This number is adjusted for the age of the population being studied to yield what is called the incidence of the disease. Another way to count breast cancer is literally to take a body count of those who die of the disease. This is also adjusted for the size and age of the population and called the mortality rate. The nature of the underlying population to which these figures are adjusted can affect the overall rate. For instance, to compare cancer rates for the U.S. population in 1970 and 1990, we first have to take into account that the population was older, on average, in 1990. Older women are more likely to develop breast cancer. More than four out of every five cases occur in women over age fifty. In order to understand whether the larger number of cases in the 1990s is chiefly due to the increased number of women who are living longer, statisticians adjust the numbers to what is called a *standard population*. This adjustment takes into account the changing number of people of differing ages at different times and is

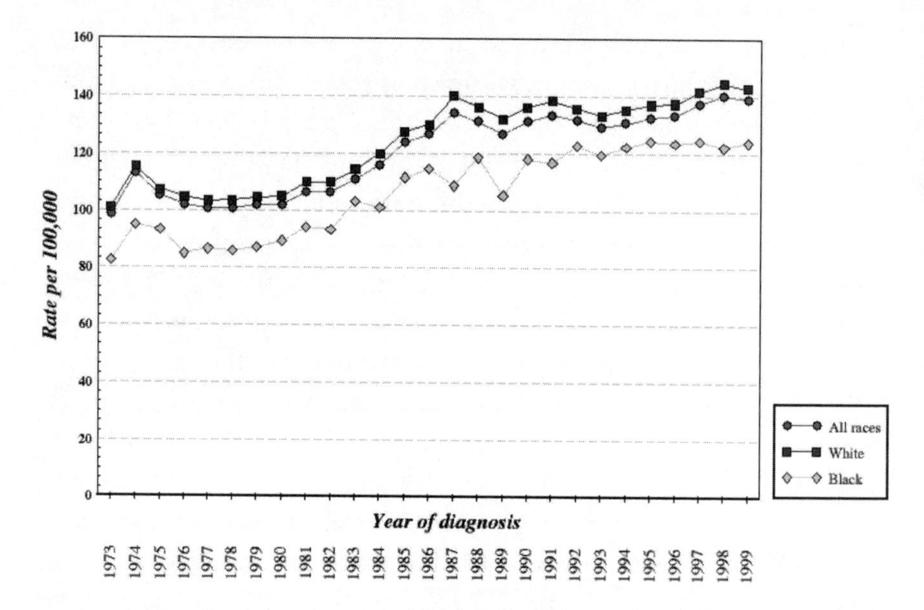

FIGURE 6.2 Breast cancer incidence time trend, adjusted to 2000 U.S. population. (Source: SEER incidence rates for individuals, available at http://seer.cancer.gov/faststats/html/inc_breast.html)

usually expressed as the number of cases or deaths occurring for every 100,000 people that year. If your mother, sister, wife, or friend contributes to the rate for any given year, adjusting her case to the population provides a sounder scientific ground for understanding the disease, but can never provide a firmer foundation for the turmoil in life with which the disease is associated.

In 1973, 100 women out of every 100,000 white women of all ages contracted breast cancer. By 1999, that number had risen to 143 (adjusted to 1970 U.S. population).[2]

Fewer African American women in 1998 developed breast cancer when compared with white women, but their rates have increased more rapidly. In 1973, 82 cases of breast cancer occurred for every 100,000 African American women. In 1999, 123 cases arose for every 100,000 black women. This 20 percent lower rate of new cases of the disease in African American women masks a very serious problem. Although relatively fewer African American women get breast cancer, a much higher proportion of them will die from it. For African American women under the age of thirty, they will develop twice as many new cases of breast cancer as will white women the same age. Nobody knows why. Nor can anyone explain why women born in the 1960s get twice as much breast cancer as did their grandmothers, or why women in areas such as Seattle, Washington, Connecticut, San Francisco's Bayview Hunters Point, Marin County, Cape Cod, Long Island, and the Great Lakes area develop the disease more frequently than women living elsewhere in the United States.[3]

By 1998, the World Health Organization confirmed what many activists had long suspected—breast cancer had become the most common cancer in women worldwide.[4] Explanations for this global pattern remain elusive. It cannot be due to more screening tests, because only a few countries routinely conduct mammographic screening, and of those that do, most focus only on women in the highest-risk group—those past menopause. In the United States and some provinces of Canada, increased efforts to look for breast cancer in white women over age forty may account for some of the growth in incidence. From 1992 to 1998, new breast cancer cases rose 1.2 percent, topping an overall jump of 40 percent in the last quarter century. In 2000, breast cancer deaths fell dramatically for white and African American women under age sixty-five in the United States and England, apparently as a result of advances in early treatment, but the incidence keeps growing.[5]

The average American woman who dies of breast cancer has her life shortened by nearly twenty years. Think about what this means. With nearly 42,000 deaths annually from breast cancer in 2001, the United States loses more than 800,000 women-years of life from the disease in a single year.[6] The toll in Europe is comparable.

Bella understood what the failure to count breast cancer and to make those counts public meant. At her Public Hearing and Education Forum on Breast Cancer and the Environment on March 2, 1993, a panel composed of members of the New York City Commission on the Status of Women heard testimony from scientific and medical experts, government agencies and community organizers regarding the role of the environment in the rising tide of breast cancer. As chair of the commission, Bella had three main objectives: to gather available information on breast cancer causes, treatments, and prevention; to bring to light the obstacles to breast cancer prevention; and to provide a basis for building policies and programs that would bring about real improvement and change.

The grand old turn-of-the-century hearing room in New York's City Hall stands three stories tall and is trimmed by imposing marble columns. We witnesses, summoned together by Bella, sat in huge leather chairs preparing for a formal public hearing on a most unusual set of topics. The only indication that this was no ordinary judicial matter came from pink placards with their black and red lettering: "Hold the Line. At One in Nine." By 2002, less than a decade later, the number was one in eight, and moving closer to one in seven in some areas.

From her position in the center of a row of high-backed mahogany chairs, Bella opened the City Hall hearing with a forceful declaration:

> We have come together to explore for the first time *ever* in a public forum of this kind, growing scientific evidence of the link between the incidence of breast cancer and environmental factors—factors such as toxic waste, electromagnetic fields, air pollution, pesticides and radiation. Our ultimate goal is to bring pressure to bear on local, state and federal public-health policy makers to broaden the parameters of the scientific inquiry so that more resources—both financial and scientific—can be redirected toward finding the causes that will lead to the prevention of a disease that is killing 44,000 women every year.

She was neither a quiet nor a foolish woman. She wanted answers.

The former fashion model Joanne Motichka—then known by the single name Matushka—prowled the astonishing scene of Bella's hearings, like a leopard stalking its prey. Angular, six feet tall, with luminescent skin and a body loved by cameras, she had stunned the public as a poster girl for the ravages of breast cancer. She produced a series of midriff-baring self-portraits revealing a thick, red scar where her small, high right breast had been, her turbaned head turned proudly to the side, nose lifted upward to display her Nefertiti-like neck. Later that year, one of her stark images appeared on the cover of the *New York Times Magazine*, headlined, "You Can't Look Away Anymore."[7]

Forcing people to see breast cancer in a gorgeous, young body turned off many, but certainly brought attention to the issue. A few years later, hiking above Thunder Bay at the First World Congress on Breast Cancer in Kingston, Ontario, Matushka told me what had brought her to this public exposure of her scarred body. When first diagnosed with breast cancer, she had not understood that her relatively small tumor could have been safely removed by a procedure called a lumpectomy, which would have left most of her breast intact. Instead, she lost her breast and her career as a fashion model. "Basically, the doctor made a mistake in my case. I lost a breast and the world gained an activist."

Matushka spoke about her childhood playing at the seashore:

> I grew up in Jersey, spent summers at the shore, with lots of pesticide spraying all the time. Everybody in my old neighborhood is getting the disease now. It is not so much that I inherited my parents' genes as that my mother and I both grew up in the same environment—in an area with plenty of chemicals all the time. I can remember my grandfather keeping gallons of stuff around the house in the garage and all those trucks spraying mists that we would run behind, pretending we were in the Twilight Zone.

Of all the battles ever waged by humankind, the effort to destroy insects is one of the most relentless. The mists that young Joanne played in contained DDT, then believed to be the perfect tool for killing mosquitoes. DDT is an organochlorine—a chemical compound made up of carbon and hydrogen to which chlorine atoms are strategically attached. Originally designed to destroy the nervous system of bugs and other pests, DDT stays in the environment for years and years; it also has a number of other problematic biological properties. In many forms of wildlife, it can function like a hormone. In humans and other animals, hormones are the messengers of the body, helping to control a variety of processes, including how they deal with stress, how big they grow, how small or large the reproductive parts become, and how well an individual develops male or female traits. Although natural hormones, such as estrogen, testosterone, and those found in plants, like soy, are short-lived and quickly excreted from the organism, agents like DDT and other chemicals that mimic hormones last a long time, especially if they are stored in fat. By the 1960s, scientists already knew that DDT was causing birds to lay eggs with shells too thin for their young to survive.

DDT first gained popularity in World War II, when it was credited with staving off a deadly epidemic of typhus in Naples, Italy, after it was directly sprayed upon more than 3 million soldiers, civilians, and children. John Wargo, professor of environmental studies at Yale University, reports that the earlier use of a group of less toxic natural compounds made from chrysanthemums, called pyrethroids, had probably already reduced the epidemic substantially. But nothing succeeds like the appearance of success. This apparent triumph of DDT in 1943 spawned extensive applications. In the Pacific theaters of World War II, entire islands were doused with the compound to reduce lethal tropical diseases, such as malaria, yellow fever, and dengue. Wargo reports that by the late 1970s, at which point some bugs had become resistant to DDT and its damaging effects on birds and other wildlife had become clear, over 5 billion pounds had been sprayed all over the world, mostly for farming.[8]

Rachel Carson made it clear why DDT should never have been used as a mainstay of agriculture. *Silent Spring* disclosed that DDT and other persistent chlorinated organic chemicals being touted for their capacity to increase crop productivity were actually poisoning our environment, sickening and killing wildlife, and probably even harming people.[9] The pesticide industry fought hard to silence Carson's voice, asking at one point, why a woman who had no children of her own should be so concerned about the future.[10] But her message forced people to look at the big picture and the full life cycle of chemicals we discharge into the environment. After Carson, it was no longer possible to measure progress simply in terms of the volume of synthetic chemicals applied to cropland or the number of insects killed. By the time she died of breast cancer in 1964, two years after her book appeared, public discussion on the issues Carson had raised was becoming much more common.

By the end of the 1970s, much of what Carson had predicted was evident. The heavy use of DDT for farming and as a spray inside homes had controlled some epidemics of malaria, but had also wreaked havoc with the environment, killing beneficial insects and birds, along with mosquitoes. Eventually, the malaria-bearing mosquitoes became resistant to DDT and the malaria parasite became resistant to the main drug used to treat it. But birds and mammals remained vulnerable to the pesticide.

Although insects have no fat, small animals that eat insects do. Fat attracts and stores organochlorine molecules in the same way that a hungry Pac-Man gobbles up dots. Fat has been called the body's natural hazardous waste site. Predators ingest and store all the pesticides from the fat of their prey. As Sandra Steingraber reminds us in *Having Faith*, her poetic exploration of pregnancy and childbirth, at the top of the human food chain lies the nursing infant, taking in her mother's milk and everything that comes with it.[11]

How long do some pesticides stay in the body? One indication comes from an unintended experiment that took place over two years, from 1991 to 1993, when a group of eight scientists lived inside a completely enclosed environment in the Arizona desert. Biosphere II was designed to test whether humans could survive on the moon or Mars by living in a self-contained bubble that relied on recycled wastes and its own internal production system for food. The physician researcher Roy Walford, who lived through this experiment, reported that after a few months, pesticide residues began to appear in the residents' urine, as the average person lost nearly 20 percent of his or her weight, much of which was fat. This release of old fat yielded a cascade of stored pesticides they had acquired years earlier.[12]

In the early 1980s, a number of studies suggested that women with higher levels of DDT metabolites in their blood had nearly four times the risk of developing breast cancer. But several later reports find no such link. What is going on? It sure looks like the bans on DDT have worked in many industrial countries. Today, in those nations in which DDT has been banned or severely restricted for nearly three decades, levels of DDT metabolites remaining in blood and breast tissue have dropped to less than one-tenth of what they once were.[13] This is good news. It also means that it is very hard to study the impact of these lower levels on breast cancer, because everybody's levels have dropped so much. In addition, there are many other types of compounds that do not stay in fat and that do affect the breast. For example, past uses of some phthalates, the agents added to plastics to make them soft, could be important causes of breast cancer today; but we have no simple way to estimate these past exposures. We know that deaths from breast cancer have declined in some age groups. We do not know, however, whether this welcome drop has anything to do with reductions in DDT.

As with many things in the environment, finding a connection between toxic chemicals and breast cancer has proved problematic. A number of legitimate scientific issues make this work quite difficult. First, organochlorines are complicated chemicals that the body processes into various forms or metabolites. Some forms of the widely used set of chemicals called PCBs (polychlorinated biphenyls) and some forms of DDT appear to goose the hormonal system. Others have just the opposite effect. If we look only at the total amounts of these chemicals and not their metabolites, we cannot know the relative amounts of hormonally active or suppressive toxins they contain. For instance, technical-grade DDT, the commercially used form that was widely sprayed around the world, contains a mixture of chemicals that vary in their breakdown time and potential to cause breast cancer. Most studies measure DDE, a common metabolite of DDT, because it is relatively easy to find, even years after exposures have ended. But, whereas DDE is much less hormonally active, technical-grade DDT, on the other hand, contains potent hormone-stimulating agents. Suzanne Snedeker, a scientist studying breast cancer and organochlorine pesticides at Cornell University, suggested that measurements of current levels of DDE are not directly relevant to the cancer process.[14] Unfortunately, no studies are under way that can resolve this matter.

In countries in which DDT use has been banned for some time, the main source of DDE comes from what we eat, not from exposure to those DDT components that may pose a far greater breast cancer threat. Thus you can have high levels of DDE but relatively low levels of the hormone-stimulating ingredients of DDT. It is entirely possible, Snedeker speculated, that studies in the United States in the late 1990s do not support a link between DDT and breast cancer, because they are focusing on the wrong metabolites, and that studies would show such a connection if they were conducted in countries with active spraying of technical-grade DDT.

So why was Matushka running in a mist of DDT in the late 1950s, singing that scary theme music from *The Twilight Zone*? At the time, spraying was frequent because officials believed that mosquitoes carried the crippling polio virus. Painted onto screens in a form that regularly released small amounts of DDT as a gas, and applied as a powder or spray on farm animals, children, and soldiers, DDT was widely hailed as safe. Some studies in the 1990s have found that a single injection of some of the chemical by-products of DDT given to rats early in pregnancy can cause male offspring to develop nipples. This would be akin to a human male developing ovaries. Other studies have found that small amounts of DDT can disrupt the way cells grow, distort amounts of hormones produced, rip cells from their anchors, and prompt uncontrolled cell proliferation.

An impressive number of studies in animals and humans show that a variety of chlorinated organic materials disturb normal hormones, even if they have not been shown yet to induce breast cancer in humans. Louis Guillette, a wildlife biologist at Florida State University in Tallahassee, showed that in 1980, after a major spill of a pesticide that included DDT as a so-called inert or nonactive ingredient in Lake Apopka, Florida, most of the alligators found in the area appeared to be female. Because they are at the top of the lake's food chain and because they have some fat into which toxins can be deposited, the twelve-foot-long alligators provide a kind of natural integrator of toxins. As a result of thinned eggs, not too many alligator babies were hatched for several years after the 1980 spill. The males that were born tended to have very small penises, and both male and female alligators had elevated levels of estrogen, the female hormone, and lower levels of testosterone, the male hormone.[15] Reports from 1999 have shown that similar problems have now occurred in other lakes, including the massive 467,000-acre Lake Okeechobee. A study published in 2002 found that frogs exposed to atrazine—one of the most commonly used fungicides in the United States—have shown up with both ovaries and testes.[16] All this is a really big problem for frogs and alligators, but it could be an even bigger problem for us.

DDT and other long-lasting organochlorines are not by any means the whole story. A number of compounds known to induce mammary tumors in rodents—what would be breast cancer in humans—do not leave any traces at all minutes after exposures have taken place. These compounds include such widely used materials as benzene and butadiene—found in gasoline and car exhausts—some plastics and more than forty other chemicals tested by the U.S. National Toxicology Program. We know a lot about these compounds, but there is also much that we may never know, because studies on these widely used toxins have not always been conducted in full light.

The City Hall breast cancer hearing was a landmark event because it brought together scientific and medical experts from a variety of fields related to breast cancer in hopes of painting a complete and clear picture of what was then known about the disease. The focus was on the link between the development of breast cancer and environmental contaminants, including exposure to hazardous chemicals, pesticides and organochlorines, radiation, and electromagnetic fields. Bella Abzug knew it was important to improve access to care and diagnosis, to devise better treatments, and to give people something to hope for. But she was tired of getting the run-around when it came to explaining why so many women who got breast cancer had so few of the so-called risk factors. Why was the generation that had started out as Rosie the Riveter working in the newly industrializing workforce during World War II and had ended up living out the feminine mystique—with lots of children and keeping impossibly clean homes with newly created compounds—developing more cases of breast cancer? Based on all that we then knew about the disease, the mothers of the baby boom generation should have been developing lower rates of breast cancer. In fact, rates were growing. So were our frustrations.

In less than half of all cases of breast cancer have any known risk factors been identified, and most of these have been tied with hormones. Risk factors are nothing but conditions or exposures that occur more often in those who develop a disease than in those who don't. Among the known risk factors of breast cancer are having a mother or sister with the disease before age thirty, getting a first menstrual period before age twelve, entering menopause after age fifty-five, having x-ray exposures especially during or just prior to adolescence, having no children or not having breast-fed one's children, having a high income, being overweight, lacking exercise, drinking too much alcohol, and getting too little sunlight and fiber.[17]

At first blush these features seem to have little in common. But they provide fascinating clues about what may be contributing to the increased risk of breast cancer. The earlier in life a girl begins her menstrual cycle and the later in life a woman ends menstruation, the greater her lifetime exposure to estrogen and other hormones. Women who take extra estrogen in the form of birth control pills or hormone replacement therapy for five years or more also have increased risks of breast cancer. Women who have used both oral contraception and hormone replacement drugs have an even higher risk than those who have used only one of these.[18] Since the end of the nineteenth century, some scientists have known that the simplest way to prevent breast cancer in mice was to take out their ovaries.[19] Yet we have given women more of these same compounds, synthetically prolonging their menstrual cycles well beyond any semblance of what was natural.

Drawing on work completed decades earlier, Jennifer Kelsey and Leslie Bernstein of the University of Southern California pointed out that most of the established risk factors for breast cancer have one thing in common: They are all tied with greater exposure to hormones. The greater the total amount of hormones in a woman's lifetime, the bigger her risk of developing breast cancer.[20] Expanding on this work, my colleagues and I developed the theory that there are two basic paths to breast cancer. One path can involve natural or synthetic forces that affect a woman's total lifetime exposure to hormones. The other path can lead to damage to the genetic material with which we are born.

The first path—the environmental one—also involves genes. Genes can be thought of as the first draft of blueprints for our bodies. They dictate what color our eyes and hair will be, the likelihood that we will go bald, and how tall we may become. Genes also regulate individual cells, telling them when to grow, when and how to build new proteins, where to go, and even when to die. For cancer to happen, genes have to lose their ability to keep cells under control. This loss of control can come about from natural exposures to agents such as sunlight or from exposures to radiation and chemicals that unleash free radicals. Fewer than one in ten cases of breast cancer occurs in women who have inherited genetic defects. This means that most cases of cancer are not born, but made. The potential for cancer to develop depends on two factors: how fast cells grow and what and how much they are exposed to.[21]

Our bodies are composed of billions of cells that grow at different rates at different times of life. One of the first steps in cell growth is the copying of the cell's DNA. Any typist can tell you that the faster you try to copy a large document, the more mistakes you make. Cells face the same problem. Rapidly growing cells, such as those of the prenatal or adolescent breast, can potentially acquire more mistakes, or typos in the genetic code, and have less opportunity to repair them before they have to divide again. If these typos change the code from "breast cells divide every once in a while" to "breast cells divide all the time," we've got problems. Even with ionizing radiation—the only known cause of breast cancer—the same dose can cause greater damage to younger breasts than to older ones.[22] When healthy genes lose their ability to keep cells under control through damage from environmental agents or from the accumulation of too many hormones at the wrong times of life, this increases the chances that breast cancer will develop.

Things that happen to women early in life appear to be especially important in determining the chances that breast cancer will occur. How do we know this? Women born to mothers who had unusually low levels of estrogen during pregnancy have less than half as much breast cancer compared to those who underwent normal pregnancy. In contrast, some studies have found that those who experienced higher levels of hormones prenatally or neonatally have nearly four times more breast cancer than women whose prenatal hormones were at average levels.[23] Something happens to the prenatal structures called breast buds when they are saturated with hormones early in life that causes them to more easily grow out of control later on in life, when other exposures kick in. The timing of exposure and the total dose of hormones involved may also be critically important. Throughout life, starting in the first trimester, the cells that will eventually become the breast continue to develop, responding to a complex exchange of estrogen, progesterone, and other growth factors. We know that girls born as fraternal twins have greater risks of breast cancer.[24] This may be because mothers of twins have more of everything, including hormones and blood, so their twin daughters get an extra boost of exposure while in the womb. We also know that girls born with a higher birth weight and those born prematurely have a higher risk of breast cancer later in life. Again, both these conditions are associated with higher levels of prenatal hormones.

Total hormonal levels and the time of exposure to hormone disturbing materials clearly can play important roles in breast cancer and other diseases. The link between hormones and genes is hardly simple. Detailed laboratory studies from the Fred Hutchinson Cancer Center, in Seattle, show that some hormonal agents actually have the capacity to change the shape of DNA, bond directly with it, or alter how DNA does its job.[25] One study found that women with breast cancer have specific chemical agents bound to their DNA.

What about environmental agents? Could they damage healthy genes or affect the amount and types of hormones that regulate practically every normal bodily function? Hormones can be thought of like keys that fit into locks in many different organs. New studies show that various environmental agents can fit into the same locks. Because the body cannot tell whether the keys are natural or artificial, cells sometimes grow in unexpected ways.

In the early 1990s, the binding of these artificial keys to the body's locks was beautifully, if unintentionally, demonstrated when cells in a Tufts University laboratory appeared to go haywire. Intrigued by the cancer process, Ana Soto and Carlos Sonnenschein had spent years painstakingly charting the distinctions between normal and abnormal growth for breast cancer cells. They were baffled when a standard line of cancer cells, named MCF-7, more than doubled in growth when left alone overnight in the lab. After an intense search, they found out what was making these cells take off. Very tiny amounts of hormones—equivalent to a single drop of water in an Olympic-size swimming pool—were leaching from the plastic laboratory dishes in which the cells sat and revving cell growth.[26]

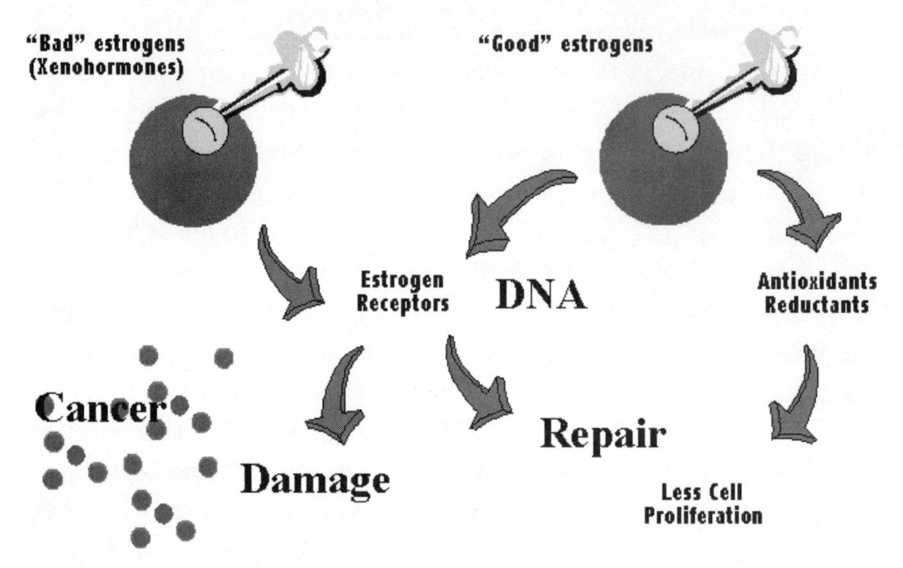

FIGURE 6.3 Both "good" and "bad" estrogens can bond with the body's estrogen receptors. Good estrogens help cells repair themselves and protect against cancer. Bad estrogens cause out-of-control cell growth and damage DNA, sometimes leading to cancer.

Normal cells have an order and a discipline that can be enchanting. As individuals we are all unique, but our healthy cells all look quite similar when magnified. Cancer cells, by contrast, are disarrayed and chaotic. They erupt and explode in a snarled, jumbled and menacing fashion. Anything that can cause cells that are already cancerous to grow faster is not a good thing. A group at Cornell University's Strang Cancer Prevention Center with whom I have collaborated since the 1990s has found that some of the body's own natural hormones have dramatic effects on cell growth. Working with radiolabeled materials that trace where materials end up in the body and can be used to measure cell proliferation, we have found that there are good and bad hormones.[27] When the body processes estrogen, it inserts an atom of oxygen connected to one of hydrogen (called a hydroxyl radical) to one of several different places on the molecule. Depending on where these hydroxyl radicals end up, they can have either positive or negative impacts. A "good estrogen" is 2-hydroxyestrone, which helps cells fix themselves and hinders cancer. In contrast,

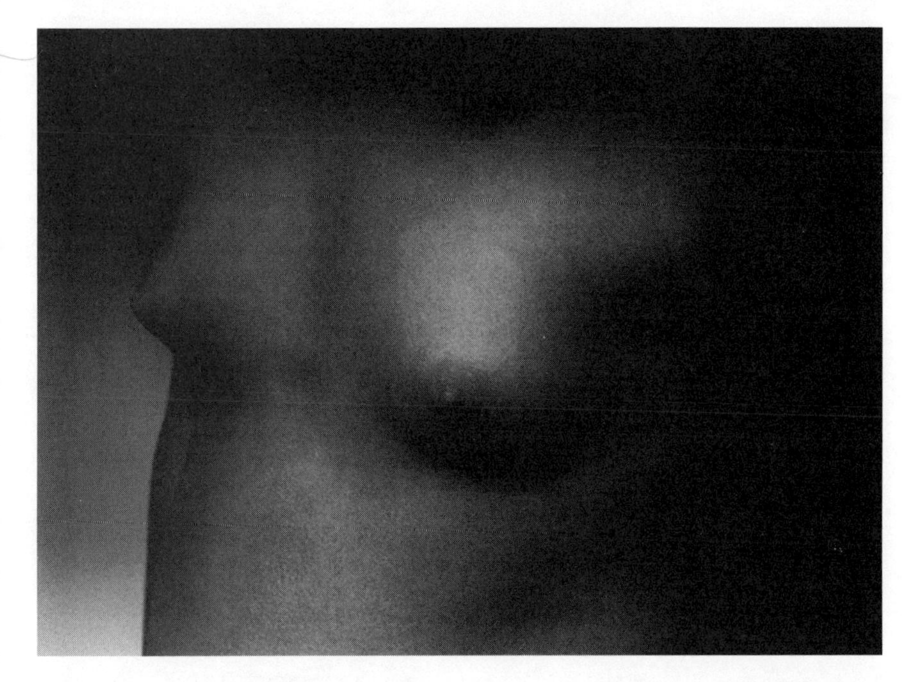

FIGURE 6.4 Breast development on a 24-month-old girl. An epidemic of premature breast growth in Puerto Rico remains unexplained. One study has found much higher residues of some plastics in the blood of young girls with breast growth. Permission I. Colon and *Environmental Health Perspectives*.

a bad estrogen is 16-hydroxyestrone, which can increase breast cell growth and hamper the capacity of cells to get repaired. Another metabolite, which picks up the hydroxyl radical at the carbon 4 position, 4-hydroxyestrone, also appears highly carcinogenic.[28]

Both good and bad estrogens can bind with the estrogen receptor—much like a lock into which several different keys might fit—but they seem able to turn the lock in different directions. Some estrogens, like those found in broccoli, tend to form good hormones that help keep cells in check and under control. Others, like those in pesticides such as DDT or some plastics, tend to form bad hormones, turn on deranged growth, and cause cells to slip their anchors, much like a ship in an unprotected cove during a storm.

Some years ago, experimental studies from the laboratory of H. Leon Bradlow, the renowned biochemist then at Rockefeller University, had shown that the body's total production of estrogen could affect breast cancer. Bradlow had done much of the basic research establishing the distinction between good and bad estrogen—between those that reduced uncontrolled cell growth and those that stimulated it. Meeting him by chance at a scientific conference in 1992, I tried to offer a suggestion.

We both knew that some natural hormones had been found to increase cancer in animals. Bradlow had shown that the bad estrogen acted similarly to this form. I proposed that some environmental chemicals might also turn out to boost total amounts of bad estrogen as well. Having spent his entire professional life charting the intricate ways of hormone synthesis, Bradlow was not naive. There was no chance this could happen, he believed. He knew all about what turned hormones on and off, and he could derive their chemical origins. He was adamant that environmental agents could do nothing at all to affect hormones.

As pleased as I was to be talking to one of the world's leading biochemists, I was pretty discouraged. In a passing effort to keep the issue alive, I asked if he would consider running a little experiment. "Let's take just a few organochlorine compounds, like DDT, and maybe other pesticides like Kepone and atrazine. Let's see if this does anything at all to the good and bad estrogens."

As the consummate scientist with the curiosity of a young boy, Bradlow agreed to look into the matter and set up an experiment to see what types of estrogens were produced by some known cancer-causing agents. Just to humor me, he agreed to throw in a known breast-tumor-causing compound, called dimethylbenzanthracene (DMBA), for the sake of comparison. The work would be done within two weeks.

To leave no doubt about what he expected, he offered this parting shot: "You are wrong. And I will be glad to show you."

A month passed. I became discouraged, thinking that he'd decided it wasn't worth his time after all. Then, about six weeks after the conference, the phone rang. It was Bradlow. "I have some news for you." His voice contained no hint of enthusiasm.

I figured he was being polite. "So? What happened?" I asked.

"Well, let's just say you were not wrong."

"Wow!" I let out a shriek.

Based on these results and other work, Bradlow and I went on to develop a theory of how some environmental agents could be playing a major role for breast cancer. One of the compounds we knew would be important was DES, diethylstilbestrol, a growth-stimulating agent that had been widely used to fatten cows, pigs, and chickens until the mid 1970s. Although we do not know what this broad use may mean for our health, we do know that women who took the same compound for just a few weeks early in their pregnancies, thinking it would prevent miscarriages, have higher rates of breast cancer and their children have higher rates of very rare, sometimes lethal cancers and other reproductive defects.[29]

About the same time that we were preparing our paper showing that some organochlorines distorted estrogens, an unlikely source brought wind of some big changes in thinking about chemicals. The United States and Canada have a long-standing binational group, the International Joint Commission (IJC), that makes recommendations about the management of the Great Lakes region that spans both countries.

IJC had been monitoring the effects of organochlorines in these lakes for years, and its staid biennial report for 1993 had issued a stunning recommendation. In light of the more than 65,000 chemicals in commercial use and the small fraction of these that had been tested for their potential toxic effects, the group called for a fundamental change in policy. Instead of assuming safety until proven otherwise, and waiting years for more testing to be completed on individual compounds, the IJC called for a precautionary approach that looked at the total weight of the evidence then at hand, including experimental, wildlife, and human studies. The report recommended that "the input of [toxic] substances to the Great Lakes must be stopped. . . . [T]he burden of proof must shift to the proponent of the substance to show that it does not or will not cause the suspected harm, nor meet the definition of a persistent toxic substance."[30]

Within the year, the Governing Council of the American Public Health Association (APHA), representing more than 10,000 public health professionals, had endorsed this approach and unanimously urged the phase-out of most industrial uses of chlorine, excepting for water treatment and the production of pharmaceuticals. In Statement 9304, the APHA explained its position:

> Virtually all chlorinated organic [carbon-containing] compounds that have been studied exhibit at least one of a wide range of serious toxic effects such as endocrine dysfunction, developmental impairment, birth defects, reproductive dysfunction and infertility, immunosuppression, and cancer, often at extremely low doses and . . . many chlorinated organic compounds, such as methylene chloride and trichloroethylene, are recognized as significant workplace hazards.[31]

In effect, APHA urged that since it would not be possible to conduct detailed studies on all 15,000 chlorinated chemicals, as a matter of public policy, the assumption should be made that all chlorinated chemicals are dangerous and should be avoided, if possible, until evidence emerges as to their safety.

However much we lack testing information on many industrial chemicals, the situation is even worse with respect to the thousands of cosmetics and other beauty products that can be directly applied to the skin and readily absorbed. We have been flying blind for years. Though required to list ingredients, cosmetic producers need not conduct detailed safety tests on these materials. Even if the Food and Drug Administration suspects that a cosmetic product causes health problems, it can't require the manufacturer to prove that the product is safe.

One class of compounds widely used in nail polish and other cosmetics are phthalates, which reduce brittleness in nail polishes and soften numerous other cosmetics, as well as plastics generally. These plasticizing agents have been found to speed up the growth of breast cancer cells and cause a range of deformities in male and female rodents. An epidemic of premature breast growth in Puerto Rico remains unexplained.[32] But we do have one clue. Girls as young as two who have premature breast development have been found to have much higher residues of this plastic in their blood, compared with those without such abnormalities.[33]

Experimentally, phthalates appear to mimic estrogen and trick the body into early puberty. A Centers for Disease Control and Prevention (CDC) study published in 2000 found that "phthalate exposure in the U.S. is both higher and more common than previously suspected, with average levels in women of childbearing age several times above the government's safety standard."[34]

In the spring of 2000, a study from the cutting-edge, nonprofit research organization, the Silent Spring Institute, warned that hair products marketed to African Americans are formulated with estrogen to make hair less dry and brittle.[35] If these ingredients were used in a prescribed skin cream, they would be subject to fairly extensive testing and approval. This is not so with beauty products. Going to the beauty parlor is an important part of life for many young black women from an early age. Could the broad use of unregulated estrogens in hair products in the African American community have something to do with why about half of all black girls begin to develop breasts or pubic hair by the time they reach the age of eight, compared with only 15 percent of white girls? Not only are young girls at risk, but they could also be exposed prenatally when their pregnant mothers use these products. Whatever may account for why African American young girls have more premature breast growth and more breast cancer than do whites of the same age, we know that genes are not involved. Blacks and whites in the United States are more genetically similar than African Americans are to Africans. And in Africa, girls typically enter puberty much later than in the United States and have much less obesity.

A decade before Bella's 1993 City Hall hearing, the government's National Toxicology Program had tested the cancer-causing potential of four times more chemicals in one year than it did in 1993. Most cancer research remained focused on treatment. No new treatments had been developed in two decades. The City Hall hearing proved to be a watershed event, as the breast cancer movement began shifting from focusing solely on medical issues to demanding answers about why more women and their daughters and sisters were developing the disease.

Barbara Brenner, the executive director of Breast Cancer Action (BCA), chronicles how grassroots breast cancer organizations sprang up around the United States in the 1990s, demanding more money for breast cancer research, prevention, and treatment.[36] These organizations were well represented at Bella's hearing by women who were not always comfortable with speaking out, but who understood they had to get over their own reticence if things were going to change. They included leaders like the unstoppable Karen Miller of the Huntington Breast Cancer Coalition and the creative social marketer Lisa Bean of the

Women's Community Cancer Project. Bean designed her own powerful pink, red, and black protest signs, choreographed their use in a massive demonstration at Boston's City Hall on Mother's Day, and then produced stirring photos of the more than 4,000 people that made up the entire scene. Others participating in Bella's hearing were Alexandra Chapman and Lorraine Pace of the West Islip Breast Cancer Coalition, Barbara Balaban of One in Nine Long Island Breast Cancer Coalition, Amy Langer of the National Alliance of Breast Cancer Organizations, and Alice Yaker of SHARE. What Bella did for the breast cancer movement was take a ball that was already rolling and give it a fierce kick forward.

Like many other advances in environmental health and in women's activism in this country, the roots of the breast cancer movement can be traced back to California and the Boston area. In San Francisco, Elenore Pred took a page from the in-your-face AIDS activists in 1990 when she founded BCA, one of the first groups to take on the issue of the environment and women's health.[37] In 1992, BCA, then led by the late Susan Claymon and Nancy Evans, put together a surprising event—a conference on breast cancer and the environment jointly sponsored with San Francisco's Junior League. The conference featured Susan Love, the activist breast cancer surgeon, me, and others. I was flabbergasted when I walked onto the stage of the wedding cake of a building left over from the San Francisco Pan Pacific Exhibition of 1916 at the Palace of Fine Arts. A packed thousand-seat auditorium of activists raised pointed questions about the role of the environment. Yes, I got a quite useful Junior League cookbook for my presentation, but I also got a look at what women could do if they organized themselves on this issue.

Another voice on these matters also emerged in San Francisco in 1992. Andrea Martin turned the same energies she had used in helping to run one of the city's most successful restaurants and then lead Dianne Feinstein's political campaigns into creating a movement to save women's lives. Martin founded The Breast Cancer Fund (TBCF), first taking on the task of showing women how to live well with the disease and highlighting the fact that the Bay Area had among the highest rates in the world.[38] Despite never having taken an overnight hike or slept in a tent in her entire life, the tiny, fashionable Martin, a two-time breast cancer survivor by age forty-two, knew firsthand that living with breast cancer was like climbing a mountain—taking it one step at a time. More than twenty women trained to scale the highest peak in the Southern Hemisphere, Mount Aconcagua in the Argentine Andes. In each of their cities throughout the United States, these same women brought public attention to the fight against the disease. Building upon the work of other activist San Francisco organizations, they created some visually compelling messages of an ice climber scaling a frozen wall with the slogan "Nobody ever said that the fight against breast cancer would be a walk in the park."

For years people had insisted there was no real problem. If there were now more reported cases of breast cancer, it was just because doctors were doing a better job of detecting the disease. This view soon was replaced by the admission that there probably was a real increase in the disease, which was basically women's own fault. Those with the disease were depicted as having been too rich, too fat, and too self-absorbed to have had enough children early in life.

Like those in San Francisco, the women of Long Island and those in Boston were furious. Why were so many women developing breast cancer? Could it have something to do with the fact that some Long Island neighborhoods were built on potato farms that farmers had deserted because nothing would grow in the soil anymore? Did it have any connection to the plumes of industrial toxins migrating into the sandy soils of Cape Cod or pesticides regularly sprayed on lawns and gardens?

In 1990, Lorraine Pace woke up near Stony Brook, Long Island, after having her second breast cancer removed and began an effort that would change the way the world looked at the disease. She wondered whether clusters of cases arose more in culs-de-sac where sewers ran together, or where golf course runoff accumulated, or near factories, nuclear plants, or highways. Pace and her friends began a door-to-door campaign making a large, crude map of the patterns of breast cancer in Long Island. Something certainly appeared to be amiss on the South Shore, where mosquito spraying had been an accepted summer ritual and where shellfish yields were dropping.

In a time-honored tradition, Senator Alfonse D'Amato asked the federal government to conduct a study of breast cancer on Long Island in

1994. The CDC seemed well suited to take on this task. The challenge to the researchers was: Can we find something besides environment that could account for the high rate of illness in this area?

The trouble was that the team at CDC never set foot in Long Island. Instead, they estimated how the women of Long Island stacked up against the known risk factors, including their number of children and social status, and concluded that these factors "could" explain the relatively high rates in the area.[39] Although technically correct, this analysis made some major errors. Probably the biggest mistake was to assume that any verdict reached by the CDC would enjoy instant credibility with the Long Island community.

In Massachusetts, a savvy group of breast cancer activists obtained state funds for scientists to conduct an independent study. The group launched what has become a major innovator in the field, the Silent Spring Institute. The institute's early work confirms that the risks of the disease vary with where women live; known risk factors explain very little of the differences. It has begun to raise important questions: What is it about being richer that increases the risk of breast cancer? Could it be tied with greater use of chemicals for lawn services, dry cleaning, termite treatments, or gardens? So far this work has produced some intriguing results. Those from areas with higher rates of breast cancer tend to use twice as many chemicals around their homes compared with areas where incidence is lower.[40]

The women of Long Island, Cape Cod, and San Francisco all confronted relatively higher rates of the disease and understood that at best all the proven risks for breast cancer explained far less than half of all cases. Besides, these risk factors were not direct causes of the disease, but were associated with something else.[41] What was it about being of a high social class that increased the risk of breast cancer? These were questions that the CDC never asked in its paper study of Long Island.

By the late 1980s, many women who two decades earlier had led the feminist movement began to reach the age when breast cancer becomes more common. Trained to go around doors that had once been closed to them, lawyers and other professionals, such as Fran Visco from Philadelphia, Elenore Pred from San Francisco, Amy Langer from New York, and Cathy Ragovin from Boston, turned their personal struggles with cancer into the most effective medical lobbying campaign ever launched in U.S. history. The world of medicine and women's health would never be the same. Within a few years, a nationwide coalition of more than two hundred breast cancer organizations was established, called the National Breast Cancer Coalition (NBCC).[42] At a rousing pep rally and reception held in the White House in 1993, Susan Love, the charismatic breast cancer surgeon, symbolically delivered to President Clinton 2.6 million signatures demanding that basic research for breast cancer be accorded more funds. At the time, the president's mother was seriously ill with the disease that would take her life less than a year later.

In 1993 the NBCC brought together more than two hundred scientists who advised that about $400 million could be spent each year on breast cancer research. Senator Tom Harkin, who had lost his sister to breast cancer, was determined to see some of the peace dividend from the collapse of the Soviet Union and the decommissioning of military bases allocated to breast cancer. Those working to convert military funds to research on breast cancer shared a secret rallying cry—"Better boobs than bombs." The growing number of women in the military made a clear case for the Department of Defense (DOD) to be involved. Senator D'Amato added his voice to an impressively bipartisan group, arguing passionately that reducing breast cancer was in the national interest. By 2000, the DOD had awarded more than $1 billion for breast cancer research.

Bella Abzug, despite her own growing infirmities, remained profoundly skeptical that all we needed to do was spend more money on better treatment and genetic research. She was pretty sure that this was not enough. She took that gut understanding to the United Nations–sponsored Beijing Women's Conference, a gathering of some 30,000, held in September 1995. Sensing correctly that women from both developed and developing regions cared passionately about the issue of breast cancer, she had invited me to speak on this topic to more than 1,000 women from around the world.

In Beijing, one woman came up to Bella and boasted, "I am the Bella Abzug of Mongolia."

Bella loved it and chuckled.

"See? They even know me in Mongolia!"

Former President George Bush made the mistake of attacking her participation in the China meeting to reporters covering his own trip to Japan, where he was collecting a hefty lecture fee. "I feel sorry for the Chinese having Bella Abzug run around there," he told *Newsweek* magazine.

He did not know that Bella was confined to a wheelchair, ordering those of us who took turns pushing her to be snappy about it. She deadpanned a reply to Bush that also made it into *Newsweek*: "He was speaking before a fertilizer convention? How appropriate."

Like many things Bella started, the campaign to find the environmental causes of breast cancer continues without her. Today Bella's portrait can be found on a large wall located right off of Harvard Square in Cambridge, Massachusetts, at 110 Church Street. The Women's Community Cancer Project secured permission from the Cambridge Zoning Commission to use the side of a popular cinema, Loews Theater, as the site of a major mural by Be Sargent. Completed in 1998, the colorful painting features twelve women who had lost their lives to the disease by that time, including Rachel Carson, biologist and author of *Silent Spring*; Audre Lorde, black feminist poet, cancer activist, and author of *The Cancer Journals*; and Myra Sadker, advocate for gender equality in education; and nine local women ranging in age from thirty to seventy: María Luisa Alvarez, Agnes Barboza, Cindy Chin, Valerie Hinderlie, Jeanmarie Marshall, Esther Rachel Rome, Susan Shapiro, Jacqueline Shearer, and Thelma Vanderhoop Weissberg. It feels a little funny to find my small portrait on the side of this building, but I am in really grand company, as one of those who continue to press the case for environmental research and precautions, along with Bella and Susan Love. The entire piece is bordered with bright black, red, and yellow symbols of some common sources of cancer-causing agents, ranging from dirty air and water and dangerous work places, to pesticides, radiation, and toxic household products and wastes.

Across the bottom of the painting is written: "Indication of harm, not proof of harm, is our call to action." Bella would have loved this, as well.

By the time of her death in 1998, a broad coalition of organizations had taken up the call for precautionary efforts. Andrea Martin had brought together TBCF, the Massachusetts Breast Cancer Coalition, the Women's Community Cancer Project, the Susan G. Komen Foundation, the National Organization for Women (NOW), and others to launch a major national campaign in the fall of 1999 to demand funding for environmental research. Knowing better than to trifle with this power, in Massachusetts and New York, the legislatures supported funding to study environmental factors at Cornell University and spawned new institutions dedicated to the issue, such as the inventive Silent Spring Institute in Boston.[43]

Organizations such as One in Nine and the Long Island Breast Cancer Coalition and the Mid-Hudson Options Project in upstate New York have persuaded the state and federal governments to launch a multiyear, multi-million-dollar project to study why women who live near two or more chemical facilities in the region have the disease significantly more often than those who do not live in such areas. In London, the National Federation of Women's Institutes and the Women's Environmental Network, with funding from the national lottery, have launched a nationwide, grassroots effort to construct maps showing the distribution of the disease.

Just as Bella fretted, the National Cancer Institute's massive increase in funding for research on breast cancer has yet to result in any major new research on basic environmental factors, beyond the work that D'Amato extracted for Long Island. The Long Island study seems bogged down, misfocused, and unlikely to be productive. As Dr. Janette Sherman reports in her book *Life's Delicate Balance*, the Long Island study has not developed a serious analysis of historical exposures in the region. Instead, the researchers collected dirt samples in plastic bags from the yards of women with breast cancer today to see whether current residues differ in those who do and do not have the disease. This approach cannot tell us anything about early life exposures or about the role of those agents that do not persist in the environment, such as many plastics, fuels, and other volatile compounds. Although we do not have absolute proof that such exposures cause breast cancer, we at least have good evidence. In contrast, we have no reason to think there should be any connection between the chance of getting breast cancer in an adult woman and contaminants found in her backyard when she is diagnosed.[44]

In truth, looking for evidence connecting environmental exposures and breast cancer is not a simple task. Despite the tremendous growth in funds for breast cancer research, we are not likely to learn why women develop breast cancer and how the environment fits into the puzzle during our lifetimes. There are several reasons. First of all, breast cancer arises after a long latency period between exposure and when the disease can be detected. Studies that measure current levels of toxic metabolites in cancer patients have been likened to the drunkard's looking for lost keys under a street light. He lost the keys a few blocks away, goes the joke, but he's looking where the light is better. With breast cancer research, we need to learn how to search in the dark. Two critical questions must be raised: What were exposures during the critical windows of development, including early life and childhood? And what was the lifetime exposure to hormonally active compounds, possibly phthalates or atrazine, which leave no trace in the body? These key relevant exposures cannot be divined through the usual instrument of human studies, a questionnaire. It makes little sense to ask people what chemicals they have been exposed to or what foods they typically ate three decades ago. We need to create other ways to obtain this information.

Second, breast cancer is so common and the factors known to affect the disease so widespread that geographic clusters will occur just by chance. Epidemiologic studies that seek to make sense of geographic patterns are best suited to detecting very large and relatively rare problems. Breast cancer risks from the environment are likely to be neither. Remember, smoking was finally accepted as a cause of lung cancer after more than fifty studies showed that the risk of disease was much higher in those who regularly smoked than in those who did not. Even so, not all smokers get cancer. The power of a study to find a risk depends on how big the risk is and how many people are exposed to it. Small risks are difficult to prove with epidemiological tools unless millions of persons are observed over long periods.

Third, although breast cancer is the result of a complex interaction between genes and a variety of positive and negative factors throughout life, studies cannot assess everything together. Many general characteristics of people, such as how much they exercise and whether they eat diets rich in agents that reduce hormone formation, cannot be easily charted. When you add this difficulty to that of gleaning information on chemical exposures, the task becomes very tortuous indeed.

Breast cancer research is at a crossroads that may not be apparent to those in the thick of it. The millions allocated to research have produced what one editor of the *Lancet* describes as a glut of same old, same old studies. Different questions must be asked to break the logjam. For women confronting breast cancer today, the central question remains this: What avoidable factors cause nineteen out of twenty cases of the disease?

We need to ask a number of new basic research questions based on this proposed rethinking of risk factors. Are there different risks for breast cancer that arise before or after menopause?[45] Why do girls who are taller and heavier at earlier ages have a greater rate of breast cancer?[46] Could growth hormones in chickens, eggs, dairy, and meat products be involved? Why do women who nurse their children have less premenopausal breast cancer?[47] Could lactation rid the body of cancer-causing agents in the breast tissue? Why do women who exercise regularly have less breast cancer?[48] Could it be because their bodies contain lower levels of all hormones, greater amounts of good relative to bad estrogen, and proportionally less fat and contaminants? Does a diet rich in beneficial xenoestrogens such as genistein (from soy) and omega-3 fatty acids (from fish oil and flaxseed) or indole-3-carbinol (from broccoli and cauliflower) protect against breast cancer in Asian women by elevating their production of good estrogens[49] or by lowering their amount of all estrogen?[50] Would diets high in such foods or supplements prevent breast cancer? Are there levels of these compounds that could prove harmful? Is the reason that blind women have lower rates of breast cancer because they generally have higher levels of melatonin—the hormone that sighted women produce when sleeping in the dark at night and that is a natural suppressor of breast cancer cell growth?[51] Do women who work night jobs have more breast cancer because their lack of nighttime sleep means they are not able to make enough melatonin?[52]

Breast cancer advocate Nancy Evans argues that we must stop relying on the incidence of women with breast cancer as the bodies of evidence needed to prove the environmental connection. In 2002, the Breast Cancer Fund and Breast Cancer Action pulled together the work of scientists from around the world on the connection between synthetic chemicals and breast cancer. They found that several widely used classes of chemicals, including organochlorines and some phthalates, are likely to cause cancer. The introduction of their White Paper on breast cancer and the environment says, "We ignore at our peril the increasing evidence that chemicals are contributing to the rising tide of breast cancer. The obligation to understand this evidence and begin to address it through implementation of public policies that put health first rests with all of us. It is in our power to change the course we are on. Now is the time."[53]

We build bridges and buildings based on safety factors. We make them stronger than need be so that they will survive unexpected assaults. When it comes to keeping women from getting breast cancer, we know a lot about what worsens the risk. We need to make better use of this knowledge. We need to stop making women into the lab rats of this earth while we wait for more certain proof. The way we live our lives should be treated no differently from those bridges we cross every day.

As studies continue to yield confusing and often inconsistent results, physicians on the front lines of breast cancer treatment across the United States are no longer willing to wait for conclusive evidence of the environment's role in breast cancer development. Instead, these doctors, such as Deborah Axelrod and Mitchell Gaynor in New York and Raymond Y. Demurs, Jadranka Dragovic, and Paula Kim in Detroit, are educating their patients and the public about how to protect themselves now. The question of the day remains: Can we organize our lives to minimize the chance that these diseases will develop, even while we wait for better scientific information on their causes?

Bella Abzug died two years after the Beijing conference. She understood that death in many religions signals the end of physical life and the beginning of an eternal, spiritual life. In Judaism this is known as *machaya hamatim,* translated as "life everlasting." As a young girl of twelve at her father's funeral and for thirty days afterward standing in the back of the synagogue, Bella publicly said the kaddish prayer to sanctify life and honor her father, since there were no male family members to fulfill this traditional task. She was made to stand at the back of the room, because Orthodox Judaism separates men and women in their observances and obligations.

The belief in immortality forces each of us to confront the limits of what we can know in this world. Maimonaides, the distinguished medieval doctor and Jewish philosopher, once wrote, "for us to conceive of life after death, an existence necessarily free of physical traits and attributes, is as impossible as for a color-blind person to grasp the colors of a sunset."

Throughout her distinguished career, Bella maintained an interest, if an increasingly skeptical one, in questioning whether the eternality of the soul could ever be known. Late one night, after rallying the troops for NOW at the Washington Monument, she struggled with the pain of arthritis and a severely swollen arm from lymphedema. This complication of her breast cancer surgery had arisen when an ignorant young doctor put a blood pressure cuff on her arm shortly after surgery.

She asked me, "What makes you so sure there is anything after we die?"

"Bella," I replied, "what makes you so strong in the face of such pain? What forces you to spend these days struggling with us here? Why don't you just go to Florida and play mah jong? Can it be anything other than the souls of all those who have gone before you?"

"I'm not so sure about *all* those souls," she sighed. "But I know my mother is still very proud of me."

"We are not going to allow you to just disappear. I know I will see you again," I assured her. She nodded without argument. She was tired and coming to what would be the end of her life. She said she was living on fury at the time. I like to think she drew comfort from the thought that her work on this issue would continue through the thousands whom she had reached.

Barely half a year later, Bella's funeral became a celebration of her gritty determination that we no longer stand by passively and wait for science before acting to protect ourselves and our families. The young woman rabbi presiding over her funeral celebration somberly intoned, "If heaven ever were male dominated . . ." She paused and looked upward, raised her arched eyebrows with a quizzical look, and waited for the words to sink in.

". . . we know things are changing right now." That brought down the house.

Shirley MacLaine, known for her efforts to communicate across time and space with those long gone from the earth, began her remarks looking straight at the coffin. "I, of course, will speak directly to Bella."

She turned to address the casket to her left and continued. "Bella, you always said you were a feminist. But I also know that you were basically a huMANist," MacLaine stressed the middle syllable.

At that moment, the microphone, which stood several feet from MacLaine and well away from any living soul, fell off its stand. The crowd gasped.

"OK, OK, Bella! I was just kidding," MacLaine replied.

7

SAVE THE MALES

Men today are half the man their grandfathers were.
—Louis Guillette

A RECENT EDITORIAL in the *British Medical Journal* asked if men are in danger of extinction. Despite earning more money, having higher-status jobs, and holding more power in society, "men have higher mortality rates for all 15 leading causes of death and have a life expectancy about seven years shorter than women's."[1] They are also having difficulty becoming fathers. Now it looks like something is wrong with baby boys: Fewer boys are being born today than three decades ago, and more of them have undescended testes and defects in their penis. More young men are getting testicular cancer than as recently as the early 1990s, and they are developing it at younger ages.[2] Some trendy magazines have even suggested that *male health* is becoming an oxymoron.

You do not need to be Sigmund Freud to figure out why our culture has been so reluctant to accept the importance of fathers for reproduction or the increased sensitivity of the baby male. When it comes to thinking about male frailty, a kind of medical macho inhibits people's thinking. For years, reports of problems in babies' and children's health have commonly been attributed to mothers. But it takes both a male and a female for any mammal to reproduce. Fathers figure in the health of their children. Things that happen to men before they try to become fathers affect not only their ability to produce healthy children but the sex of those they do create, and their prospects of becoming grandfathers.

Moreover, within our culture, reproductive health problems, even more than other health issues, have been widely believed to be the luck of the draw. When a child enters life with serious deformities, we tend to attribute these tragedies to fate, to unavoidable factors that can never be fully sorted out, or to something Mom did wrong. We all know, of course, that those who seek to become parents should not smoke, drink alcohol, or take drugs. But engaging in bad habits cannot completely account for the more than 450,000 babies born prematurely each year, the 6,500 who die of birth defects, and the thousands of others who are born with major abnormalities.[3] Four out of every five of these misfortunes remain unexplained.[4]

This chapter reveals how the health risks that children suffer from the environment have remained in dispute and unresolved far longer than necessary. There is no denying that the science behind these issues is complex. On the other hand, those concerned with spinning information on environmental health have often put corporate interests far ahead of concerns about public health. Whenever researchers present their findings of potential hazards to male reproductive health, whether increases in birth defects or declines in the births of baby boys or sperm count, others magnify and manipulate valid scientific disputes to explain that the harm is either not real or unimportant. Thanks to these countervailing efforts, some two decades after the first reports of hazards to male reproductive health, we know less and less about more and more. The reasons have partly to do with the intricacies of the science, but also result from the skillful ways in which some in the corporate world have effectively blocked research, canceled studies, pulled funding, and employed sophisticated public relations campaigns to cast doubt on these questions. If you don't want to know, just don't ask.

Today, with one out of every five couples having difficulty becoming parents when they wish to, fertility centers in the United States and elsewhere are full of advice to mothers and fathers about how to live their lives. If you want to make a healthy baby, avoid cigarettes or pot, alcohol, cocaine, high temperatures, and intensive exercise.[5] But what about exposure to toxic chemicals? Does it make sense that recreational habits would affect sperm—as they do—but that other agents in the environment would not? Reproduction is an exquisitely complex process that begins long before the egg and sperm connect. Because both these critical components of reproduction are bathed in fat, and because fat naturally attracts some toxic chemicals that are rich in organic molecules, anything that gets into fat has a chance to be present at the moment of conception as part of the egg or sperm.[6]

Vulnerability to anything is greater the faster a cell divides and grows. This is why a developing fetus is so sensitive to toxins. The same applies to sperm, which are among the fastest-growing cells in the human body. Sperm in the adult male exceeds brain, toenails, saliva, and hair in its capacity to take in tiny errors as it develops.

Outside of good and bad personal habits and some workplace hazards, few risks to reproduction have been identified. Nutrition, exercise, the use of some medicines and recreational drugs, and family history—all the usual individual factors—are of obvious importance. But when it comes to risk factors beyond an individual's control, we know far less than we should. Plenty of couples lead exemplary personal lives, yet still face major problems conceiving children. What do we say to them? What do we say to parents whose children are born with holes in their hearts or undescended testes or other birth defects, even though their families have no history of such problems? What do we tell the growing numbers of young men who are developing testicular cancer in every industrial country today?

Our ignorance on this subject is not strictly a scientific matter. When I was working for Douglas Costle, the EPA administrator under President Carter in the summer of 1977, I was part of a group of scientists and policy analysts struggling with how to set up the agency's systems for reviewing information on toxic health risks. The pace of our work was at best plodding. But one report flew like a hot potato across the administrator's desk: A group of men in California had figured out that—in the macho vernacular of factory workers everywhere—they had all been shooting blanks. They all worked in the Agricultural Chemical Department of the Occidental Chemical plant in Lathrop, California, producing a pesticide called dibromochloropropane, or DBCP. DBCP is a compound of chlorine, bromine, and hydrocarbons that is injected into the ground to kill worms that attack bananas and other fruit. It was a kind of miracle pesticide, yielding very productive harvests, but it had unusual biological properties.

Of thirty-five married men who had worked in this California factory, not one had been able to become a father. After learning of this outbreak of sterility, the Oil, Chemical, and Atomic Workers Union requested that the company survey plant conditions. When Occidental Chemical refused to do so, independent filmmakers and the union carried out their own survey and confirmed that most of the workers had such low sperm counts that they qualified as sterile.[7] A probe of the matter quickly revealed that EPA and its predecessor agencies had had evidence sitting in their files for more than a decade showing that, in rats, DBCP shrank gonads, and caused weight loss, hair loss, and gross lesions of the lungs, kidneys and testes.

The studies that demonstrated this were conducted for Shell and Dow Chemical Company, the manufacturers of DBCP, in the 1950s. At first, both companies kept the information within their own files. In 1958, company researcher Mark A. Wolf provided Dow Chemical with an internal report of studies of various concentrations of DBCP, noting that the compound is readily absorbed through the skin and causes severe kidney damage. "The use of this material as a soil fumigant," he wrote, "should be approached with caution. Personnel handling this material should be made aware of the hazards involved."[8] In 1960, the fellow in charge of industrial hygiene at Dow, V. K. Rowe, recommended that DBCP deserved a "tough label"—meaning it should be treated as highly toxic.

Three years after experimental studies first found that exposures of DBCP to skin proved highly toxic to the testes, these findings were published by Dow toxicologist Ted Torkelson in the journal *Toxicology and Applied Pharmacology*. He noted that DBCP severely damaged the testes in rats, guinea pigs, rabbits, and monkeys at very low levels of exposure—1 part per million.[9] That same year, Torkelson wrote guidelines to the company, advising that DBCP workers should "wear full-gear face masks equipped with vapor canisters, and [their employers should] provide clothing impermeable to the material."[10]

Skin contact should be prevented completely, Torkelson advised. "Protective clothing impermeable to the material should be worn if the likelihood of skin contact exists. Standard rubber or neoprene gloves do not offer adequate protection and should not be relied upon for keeping material off the skin."[11]

None of these warnings ever appeared on the manufacturer's safety data sheet—the official source of information for the workplace—for DBCP until 1977, after reports of damage to the California factory workers had surfaced. At that point, the label warned that the compound "may be absorbed through skin," but still did not mention that the testes were the target. After evidence of testicular damage in workers bubbled to the top of EPA, DBCP became the fastest chemical ever regulated. The administrator immediately issued an order banning nearly all its uses in the United States. But it remained perfectly legal to produce DBCP as long as official Occupational Safety and Health Administration guidelines were met. Production expanded in Mexico and other developing countries and at a small California firm called AMVAC, until it was shut down after repeatedly releasing DBCP into the air and water.[12]

How was it possible that the hazards of DBCP remained hidden for so long?[13] In the area of pesticides and toxic chemicals, EPA has struggled for years to define its proper role. Despite impressive rhetoric in the Clinton administration and the involvement of people with solid environmental credentials, when it comes to the assessment of toxic materials, EPA remains chiefly an information processing agency. It must rely on the Toxic Substances Control Act (TSCA) (called by some the Toxic Substances Conversation Act because so much of it has involved talk rather than action), its much-reformed Federal Insecticide Fungicide and Rodenticide Act (FIFRA), or its Food Quality and Protection Act (FQPA). It neither conducts nor funds basic research on toxic chemicals. Instead, it reviews reports provided to it by the very companies it is supposed to regulate, as well as those produced by the National Toxicology Program. For years, no matter what part of the political spectrum sat at the top of the agency, EPA has permitted companies to evaluate and convey their own toxicity studies with little independent review. The results have not been salutary.

Early on, shoveling boxes of internal reports and raw data, uncataloged and unorganized, into the agency put companies in compliance with the law. When I began working at EPA in 1976 as executive secretary of the Administrator's Toxic Substances Advisory Committee, I visited rooms where millions of pages of industry records were stored in file boxes. Full-time staff worked at sorting, labeling, and organizing those materials, but the boxes came in faster than they could be dealt with. EPA was years behind. Thus it was never a surprise to learn that some ominous and worrisome information had technically been in the agency's possession for a long time. We referred to this dumping of boxes of data as malicious compliance.

As a result of this cumbersome system, the United States lags behind Europe and especially Scandinavia in its capacity to take action on the information that sits in its files. There is no truly independent source outside the federal government that regularly evaluates the potential hazards of chemicals to human health. And the government, as we have seen, is hardly immune from outside pressures.

A few years later, in 1985, I was working at the U.S. National Academy of Sciences. A soft-spoken Texas lawyer named Charles Siegel of the Houston law firm of Barron and Budd came to see me. He wanted to talk about an article I had published in 1981 with Harvey Babich on the sorry history of DBCP, in which we had reported that evidence of its toxic effects on reproduction had been around many years before it was banned.[14] Siegel told me that more than 20,000 men in Costa Rica had become permanently sterile from using DBCP after the United States ban had been put in place. Dow and Shell, he said, had continued to export large amounts of the compound to third-world countries. Unaware of its hazardous nature and having never seen a safety data sheet in any language, workers in Latin America would chuck the leftover chemical into streams to kill fish for their families to eat. In their field work in the hot, humid Atlantic and southern regions of Costa Rica, the men used no gloves, protective clothing or equipment of any kind to prevent direct skin contact with the pesticide. They had never heard of Torkelson's advice, and if they had, it is not at all clear that they could have acted on it. Protective clothing or headgear is hardly useful where temperatures can reach 100 degrees and humidity 100 percent.

Once the EPA banned the use of DBCP in 1980, people began to ask who knew what and when. They confirmed what unpublished laboratory animal studies conducted in the 1950s and 1960s had shown: DBCP is toxic to the male reproductive system in every species tested. A single snapshot in time, a cross-sectional study of twenty-three male workers employed in the 1970s at a DBCP production plant, found that eighteen of them had serious deficiencies of their sperm.[15] Twelve who had no viable sperm at all—a condition called azoospermia—had worked between 100 and 6,726 hours with DBCP. Six with very low sperm counts of less than 10 million sperm cells per milliliter had worked with the material from 34 to 95 hours. The remaining five men had normal sperm counts and the lowest exposures, ranging from 10 to 60 hours in total. Those lacking sperm had elevated levels of some female hormones as well. Samples taken from the testes of some of these men revealed atrophy of the internal architecture of the testes. Four years later, although some of the men had recovered, eleven remained completely sterile.[16] Another study of men working in a different production facility for eight years found that half of the twenty-five workers showed either low levels of viable sperm or none and also had elevated levels of female hormones.[17] Those with the greatest reductions in sperm count had the longest periods of exposure. Men who had worked producing DBCP for ten years had no sperm at all in their ejaculate and did not appear to recover to normal levels, even seven years after exposure had ended. Another study of a small group of workers in Israel who had been applying DBCP as a pesticide found that they also had serious difficulties becoming fathers and produced three times more daughters than would have been statistically expected.

Based on all this work, and the U.S. federal government's first proposal to ban the use of DBCP in the United States in 1977, the sterile Costa Rican workers and their spouses filed suit against Dow and Shell. These suits were thrown out of courts in California and Florida on grounds that foreign companies had no standing to sue in the United States. Siegel succeeded in bringing the case to Texas, home of Shell's headquarters and the largest Dow-owned chemical plant in the United States. The suit was dismissed in the Houston State District Court. The Appeals Court overruled the District Court. The Texas Supreme Court agreed, declaring that Dow, Shell, and other corporations could not apply different health and safety standards to products and processes they ship overseas. In 1991 the case was settled against all the defendants for nearly $20 million. Shortly afterward, Texas changed its law to deny foreigners the ability to sue U.S. companies.

Siegel spent his entire legal career of more than fifteen years pulling together this case, which eventually consisted of 13,000 additional men from Honduras, Nicaragua, Guatemala, Panama, Ivory Coast, Burkina Faso, Philippines, Ecuador, Saint Lucia, Saint Vincent, and Dominica—all of whom had worked on banana or pineapple plantations. The final settlement rewarded those with the lowest sperm counts the most funds and was settled in 1997 and 1998 with all the defendants except Dole for more than $50 million.

To assist Siegel, I began an analysis of the sex of the children born to those DBCP workers who were able to become fathers, but I was never able to complete it. Even if we had collected all the data, we faced some major cultural problems in determining whether unusual patterns had occurred. Because becoming a father is so important in many of these countries, we could never be sure who really had fathered children and who had not. But given what we know from animal research and workplace studies showing reduced numbers of baby boys fathered by workers with heavy exposure to DBCP, the capacity of these workers to become parents was unlikely to have been normal. Because the cases were settled and sealed, and because no governments or independent groups have ever even asked, we will never know. The losses to all those families are incalculable, as are the losses to the science of environmental health.

At about the same time that reports of the effects of DBCP were racing through EPA, a massive explosion destroyed a small factory located in the Po River valley town of Seveso, north of Milan, near the beautiful lake region of northern Italy. A loud, screeching, whistling sound erupted shortly after noon on July 10, 1976, as several hundred pounds of dioxins were released into the air over the town. In his account of this disaster in *The Poison That Fell From the Sky*, John Fuller described how

Seveso became a real-world example of what Rachel Carson had only conjured—a place rendered silent by poisons.[18]

Within a few hours of that weekend eruption, hundreds became nauseated and sickened, as a white film settled over the area. Skin lesions appeared on some children. Not until two days later, when cows, dogs, chickens, cats, and birds began to keel over, did authorities begin an investigation. Within two weeks of the blast, a chemist deduced that because of characteristic lesions and the way the animals swelled up before dying, dioxin was the culprit. Within three weeks, some 736 people living closest to the plant were evacuated. Chloracne, a severe and painful skin disorder usually associated with dioxin, broke out on some of those who had been most exposed. Eighty young children were moved to another location as decontamination efforts began.

About 37,000 people were thought to have been exposed to the chemical cloud. Approximately 80,000 local farm animals were slaughtered to prevent contamination from moving up the food chain. Zones of exclusion were set up, with the most contaminated area, Zone A, covering 110 hectares. This was completely evacuated and later became a park—Seveso Oak Forest. In the two next most contaminated zones, B and R, farming and the consumption of local agricultural goods and meats were strictly forbidden.[19]

Recognizing that an unprecedented environmental catastrophe had taken place, Pietro Mocarelli, a physician researcher at the local hospital, and his colleagues drew and stored blood from those who had been exposed as they filed in for help over the next few weeks. At the time of the accident, nobody knew how to measure dioxin in blood very accurately. That would come nearly a decade later, in 1987. The Seveso incident became a chance experiment about how dioxin affects health.

As soon as local physicians understood that the toxic cloud had contained one of the most damaging agents then known, they began to advise those who were pregnant that they faced potentially grave risks. Though abortion was then illegal, the local health authorities granted special exemptions to any pregnant woman who chose to end her pregnancy.[20] Most of the young Catholic women who were pregnant at the time of the explosion traveled by bus over those beautiful, serpentine, high mountain passes of northern Italy to medical clinics in Swiss towns, where they legally ended their pregnancies. As a mother of two much-wanted children, I cannot begin to imagine their heartbreak.[21]

It later became clear that these sad journeys had made sense. Throughout pregnancy, the mother's body becomes a fortress designed to do whatever it takes to protect her offspring from the time of conception until birth.[22] The placenta acts as a living shield, taking what it needs, and sometimes what it simply cannot refuse, from the mother's body.[23] *Remarkably, when the discarded placentas from the women of Seveso were examined, half proved to be abnormal.*[24]

An equally important part of what makes for successful reproduction is the umbilical, or maternal, chord, through which nutrients, gases, toxins, or waste products enter or exit the placenta. This chord can be 4½ to 5½ feet long—longer sometimes than the mother is tall. The cells in the maternal chords from the Seveso women whose pregnancies ended were examined. Normal cells follow rules and stay in line. Aberrant cells are defined as those that are determined not to go where they were supposed to—these are the troublemakers of our bodies. Usually, our healthy suppressor genes, or repair enzymes, are called on to fix aberrant cells or to make them die, when they are not fixable. That is how we survive assaults from ordinary life, like being exposed to too much sunlight or taking in too much alcohol or other easily abused foods or drugs. The chord blood of those who chose to end their pregnancies in Seveso had many more aberrant cells than those found in women outside of the region.[25]

The effects of the disaster were not limited to the ended pregnancies of these women or to short-term skin lesions, headaches, and teary eyes. A total of 181 families were evacuated from the most heavily contaminated area after the accident, while people living in the second-most contaminated area (zone B) were asked to follow strict health regulations. Even though the number of people exposed was very small, in 1993 Alberto Bertazzi of the University of Milan reported rising rates of soft-tissue sarcoma, non-Hodgkin's lymphoma and Hodgkin's diseases, and leukemia. These diseases typically take years to develop.[26] By 1997, deaths from leukemia, Hodgkin's disease, and multiple myeloma were three to six times higher for persons who lived in the most contaminated zone, compared to those outside this area. Liver cancer and lymphoma rates were highest in those who had lived in these contaminated zones the longest.[27]

A more troubling finding from Seveso is a deficit in the number of baby boys born after this episode. Men who were young at the time of the explosion have continued to father fewer boys, even when married to women from outside the affected areas. Between April 1977 and December 1984 (the period corresponding to dioxin's half-life, or the amount of time it takes for half of the body's dioxin to be naturally eliminated), 74 children were born to parents in the zone of greatest exposure. Of these, 48 were female and 26 male. The stored serum samples taken from the exposed populations in 1976 were looked at with a new technology in 1984 that could measure dioxin more precisely. Not a single baby boy was born to the nine couples with dioxin levels above 100 parts per trillion. Since 1985, the proportion of male babies born in this population has returned to normal levels and overall fertility has increased.[28]

For the people most heavily exposed, the problems have persisted. Scientists believe that in any normal population usually about 106 baby boys are born for every 100 girls. The younger a boy was when first exposed to dioxin in Seveso, the lower his chances of producing any male children. Men who had been exposed to dioxin, married to women without such exposure, fathered a total of 88 boys and 103 girls. Where both parents had been exposed, they produced 113 boys and 137 girls. For women exposed to dioxin, with husbands without such exposure, there was no evidence of reduction in male babies; they produced 127 boys and 106 girls.[29]

Figuring out what might distort the normal ratio of male to female births—called the sex ratio—in a single area is not a simple task. Several analysts have asked whether the same environmental exposures that are linked to sex-ratio alterations in the children of highly exposed workers or those of Seveso's fathers might also cause changes in the sex of children conceived in other neighborhoods with similar exposures, or in populations where such exposures are widespread.

Why don't we know more about what has affected the sex of our children or their reproductive health? What are we up against when it comes to sorting through what is going on with male reproductive health? Theo Colburn, Diane Dumanoski, and Pete Myers's important book *Our Stolen Future* appeared in 1996, arguing that commonly occurring hormone-disrupting agents lay behind some defects in human reproduction and in wildlife. A stream of critics issued a series of coordinated critiques of its charges. The Chemical Manufacturers Association directed journalists to several experts it had polled in anticipation of the publication of this book.[30] The experts included infertility specialists like Harry Fisch of Columbia University and Larry Lipshultz of Baylor College of Medicine. The Chlorine Chemistry Council provided its own Web site of criticism of the book, taking studies conducted by eminent researchers that supported the council's position that the capacity of chlorinated compounds to impede reproduction remained unproven.[31] The authors of the book in response developed their own site of information supporting their positions and responding to the steady stream of criticism coordinated by industry.[32] They noted ecological reports of hermaphroditic polar bears, heavily contaminated fish, and other wildlife with mixed genitalia, along with reports of various indications of harm to human male reproductive health.

The matter of whether sperm counts had dropped was subject to much heated discussion. In the 1980s Niels Skakkebaek, a Danish fertility specialist at Copenhagen's Righospitalet, wanted to figure out why it was becoming so difficult to find good sperm donors. He began studying men not known to be working in hazardous jobs and was shocked to see that nearly all of them had sperm counts half of what was considered normal. He learned of similar problems and findings in a number of other countries, including England and the United States. In an effort to see whether this was a broader phenomenon, Skakkebaek, Elizabeth Carlsen, and other colleagues combined data from sixty-one studies conducted from 1938 to 1991 at different periods, finding that male fertility had dropped by about half in four decades, from 1,130,000 sperm cells per milliliter to 660,000.[33]

This analysis was criticized for having included measurements of sperm from men visiting fertility clinics and those from widely different countries with different baseline rates of sperm. Some critics argued that the reported reductions in sperm count and increased numbers of defects in young human males were not taking place. They indicated that sperm counts in Southern California were much lower than those in

New York, and that the former had been measured later, creating an artificial downward trend so that no general conclusion could be drawn. This ignored the fact that California's population includes much higher numbers of Asian and Hispanic men, who are known to have lower sperm counts.[34] Further, these critics argued that if there were increased numbers of sterile wildlife, fish with mixed sex traits, or experimental studies showing damaging effects of endocrine-disrupting agents in the laboratory, these were just not relevant to humans. Their bottom-line position was: We need more research. They are correct. We always need more research. But what do we tell the families of children born with birth defects in the meantime?

We can tell them that research is under way. But think about this. In 1983, the National Academy of Sciences issued a report—one of the first I worked on—showing that toxicity testing information had been developed on fewer than 10 percent of the 75,000 chemicals then widely used in commerce. We called for more studies on the 3,000 chemicals then in use in highest volume. In response, the chemical industry announced a voluntary program intended principally to put off government actions.[35] In 1998, the Environmental Defense Fund, the Chemical Manufacturers Association, and EPA agreed to develop a High Production Volume Challenge Program to provide basic toxicity information in the form of "SIDS," or screening information data sets.

What had gone on in the intervening two decades for toxicity testing? Not enough. Under the current program, with its voluntary and mandatory components, a chemical manufacturer, or a group of them acting as a consortium, can sponsor a chemical for testing and take it out of potential government testing. So, some two decades after passage of the first federal laws regulating toxic chemicals, emissions of toxics have dropped[36] and more testing appears under way. Less independent auditing of the effort is available. Researchers continue to study compounds one at a time, and squabbles about metabolites and mechanisms have become increasingly complex. And studies address the ordinary, but complex, combined exposures that take place every day.

About this ponderous process of developing toxicity test information, Carl Cranor, professor of philosophy at University of California Riverside, notes a dark side. Because we lack testing information on most of the toxic chemicals widely used today and because we study them one at a time, waiting for these findings to be developed before acting to control suspect compounds effectively permits uses and exposures to continue until sure evidence of harm is at hand. The question of what should be done while we wait for toxicity testing can be thought of in terms of what economists call the question of the downside risk. Would you rather err on the side of overcontrolling a safe agent or on the side of undercontrolling a hazardous one? Cranor argues, and I agree, that as a matter of justice in a free society, we should prefer to accept the risk of mistakenly controlling a safer substance over the chance of falsely allowing a dangerous one to be widely used.[37] Basically, afflictions resulting from avoidable exposures in the environment are unjust because they impose unfair and inequitable burdens on those who are subject to them.

Justice and equity are hardly on the minds of those couples who are unable to become parents, those who cannot produce male offspring, or those who struggle to raise children born with defects. The scientific literature has reported increases in disorders of the male reproductive tract, including split penises (hypospadias), undescended testes (cryptorchidism), testicular cancer, and male infertility. Declines in testicular volume have also been reported. Moreover, intriguing connections have been found between these disorders and exposure to hormonally active agents, such as some plastics, fuels, and pesticides.[38]

The ultimate birth defect—death—has been tied with previous uses of DDT in the United States. As we saw in the last chapter, lower levels found today make it quite difficult to study whether past exposures have played any role for breast cancer and other diseases that can take a long time to develop. Knowing that DDT and its metabolites have subtle effects on growth, Matthew Longnecker and colleagues at the National Institute of Environmental Health Sciences took a creative look back at forty-year-old stored blood samples and birth records from more than 55,000 mothers and babies. In a study published in *Lancet* in 2001, they determined that mothers with the highest levels of pesticides in their blood had four times the chance of losing a child at birth. They believe that this could explain a past epidemic of preterm births and account for as much as 15 percent of all infant deaths in the 1960s.[39] Where in the United States were the highest amounts of DDT ever used? We cannot be sure, but we know that poor, black southern regions, which also have had among the highest rates of infant deaths, are among those with the greatest use. More than twenty-five countries today continue to use DDT.

Recent studies have also confirmed something that Lave and Seskin showed in 1970. While DDT may no longer be contributing to infant deaths in the United States, air pollution appears to be contributing even now. A study by Mildred Maisonet and researchers from the Centers for Disease Control of all births from 1994 to 1996 in six northeastern cities—Boston, Massachusetts; Hartford, Connecticut; Philadelphia, Pennsylvania; Pittsburgh, Pennsylvania; Springfield, Massachusetts; and Washington, D.C.—found smaller babies born in regions where air pollution is higher.[40] We know that babies who weigh less at birth go on to have a host of other health problems. In Europe, one large study of young adolescents who lived in dense urban areas with higher levels of lead and volatile materials, such as those released by cars and factories, reported that these children had smaller testes along with reduced kidney function; those with similar backgrounds from rural areas did not suffer from such problems.[41]

When it comes to sorting through what is going on to affect the health of young children, and why more of them appear to be born with reproductive defects, the early hormonal environment appears profoundly important. A CDC report in the United States showed that the rate of baby boys born with split penises had nearly doubled from 1970 to 1993. Increased rates of these and other birth defects in baby boys have been reported in ten countries.

During this same period, 1970 to 1993, the rate of testicular cancer grew by at least 50 percent in most industrial countries. The evidence suggests that the total amounts of estrogen to which a boy is exposed prenatally can play a role in whether he develops this disease. Boys who are born as twins with a sister have been found to have a higher rate of testicular cancer later in life. We know that mothers of twins have more of everything, including estrogen. Perhaps this higher-estrogen prenatal environment gives rise to the later growth of testicular cancer. We also know that boys born to mothers who took the synthetic estrogen diethylstilbestrol when pregnant tend to have serious reproductive difficulties, including possibly increased testicular cancer and sterility. Work-

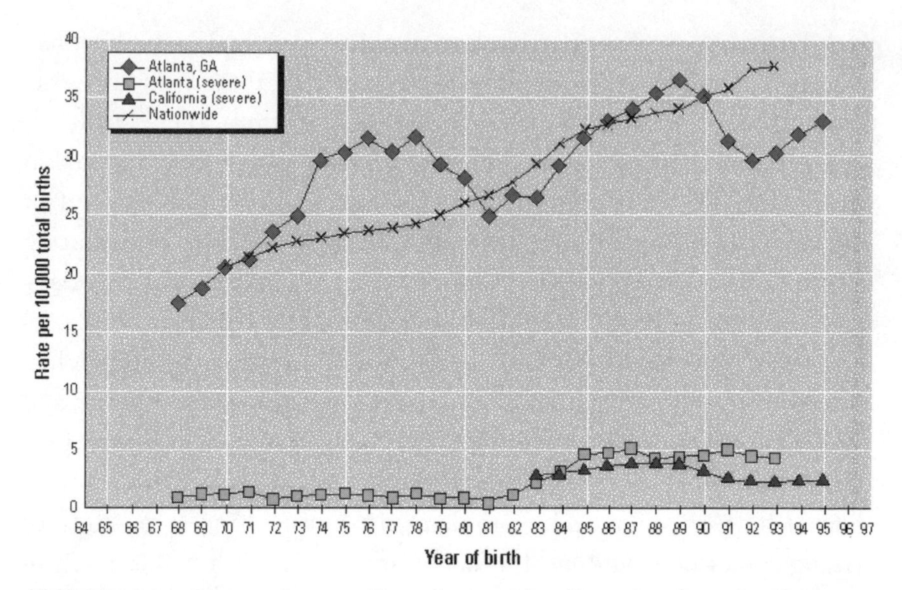

FIGURE 7.1 Changes in rate of boys born with split penises from the 1960s through 1990s. (Source: Paulozzi, *Environmental Health Perspectives* 107, no. 4, April 1999).

place hazards to which men are exposed also affect the sex of their children, as well as their health when they become adults. A number of studies have shown that men who work regularly with polyvinyl chloride or with degreasing agents, like trichloroethylene, have increased risks of testicular cancer as well as impaired fertility.[42]

Since 1960, the incidence of testicular cancer has doubled or more than doubled in each of the four Nordic countries. During the same period, testicular cancer also increased in England, Wales, Scotland, Australia, New Zealand, Slovenia, Poland, Spain, Colombia, Japan, India, and the United States.[43] The increases were consistent in all populations, were of similar size, and were reported by long-established cancer registries.

A study of male reproductive disorders in four Nordic countries found that while rates of testicular cancer have been rising, the quality of sperm has declined.[44] Sperm quality and quantity have also dropped in the United States and other European countries.[45] All these defects in male reproductive health are likely to be related to one another. A number of studies have found that men who have had undescended testes as boys tend to have higher rates of testicular cancer as young men. Men whose fathers have had reproductive difficulties also tend to have higher rates of testicular cancer. And men who have had testicular cancer also tend to produce fewer boy babies.

Also over this same time, autopsies conducted on young men who had died in car crashes and other accidents in France and Finland have found that the size of their testes has declined. An award-winning Japanese television documentary, "Hormone Havoc," disclosed similar results for Japan. Because of unresolved political and cultural issues, however, these findings are unlikely to ever appear in the scientific literature. Men who work with solvents like ethylene glycol ethers were found in the mid-1980s to have difficulties becoming fathers. They were also shown to have smaller testes than those without such exposures.[46] A 2002 study of men being evaluated at a fertility clinic found that those with lower sperm counts also tended to have distortions in sperm shape and speed, and higher blood levels of PCBs, DDE, and other pesticide residues compared with those whose sperm appeared normal.[47]

The importance of what happens to mothers and fathers before they become parents can be seen from an original piece of research done in the Spanish province of Granada, known to use the highest amounts of pesticide in the country. Although this region's fruit and vegetable crops take up only 4.65 percent of the country's farmland, they are treated with 61 percent of the pesticides used in Spain. Along much of the Mediterranean coast, greenhouse crop farming is so widespread that the plastic sheeting with which these greenhouses are covered can be seen in space photographs as concentrations of blue dots. In these enclosed greenhouses, workers regularly use high levels of organophosphates and some chlorinated pesticides that are banned in the United States because they have been shown to distort the human body's production of hormones. A group of researchers led by Jose Garcia-Rodriguez, from the University of Granada, compared rates of surgical corrections for undescended testes across different regions of Granada.[48] In the districts where pesticide use was highest, the procedure was more than twice as common per capita as in the lowest-use areas.

Could all these findings be mere coincidence? Could there be a connection between fat-seeking organochlorine contaminants and underscended testes in young boys? A recent study looked at residues of twenty-six different organochlorines in the fat of boys going through surgery to correct their disorder, compared to boys with normal testes. They found that young boys with undescended testes had much higher levels of two compounds that are pesticide metabolites—hexochloroepoxide (HCE) and hexachlorobenzene (HCB).[49] Left unanswered by this work is what the prenatal exposures of these boys could have been, what agents their fathers and mothers were exposed to prior to their conception, and whether all this played any role in producing this defect as well.

Further evidence of a link between male reproductive health and pesticide use has been generated by a team working with Vincent Garry of the University of Minnesota. They have found that boys born to men who worked as pesticide applicators have much higher rates of birth defects.[50] In agricultural areas, increased rates of birth defects also appear in the general population. For example, in western Minnesota, where spring wheat is a major crop and fungicides are sprayed by air, 2.6 percent of live births showed anomalies, as compared with 1.83 percent of births in less heavily agricultural regions. The increase was especially pronounced for infants conceived in the spring, when chlorophenoxy herbicides, like those that were made in Seveso, were routinely applied. Interestingly, more male than female infants were affected by birth defects.

Both human and animal evidence shows that *when* exposure takes place can be even more important than the *amount* of exposure. Prenatal exposures weigh more heavily than any others on male reproductive health. Disruptions that take place when the reproductive organs first start to form can create chemical imprints for later developmental disorders ranging from undescended testes, penile defects, and testicular cancer to reproductive difficulties.

To see how these disorders are connected, you need to understand a few basic facts of biology. For the first two months of embryonic life, we all have unisex sex organs. As the eighteenth-century French philosopher Diderot noted, whether a person becomes male or female depends on whether the balls stay inside the body or drop outside.

At the moment of conception, all embryos are destined to be female, but between the sixth and ninth week, the Y chromosome, found only in boys, sends out signals that trigger male development. Specialized "boy-making cells," called Sertoli cells, form and begin to direct the development of masculine traits. Once this happens, the sex of the fetus is pretty much set. The Sertoli cells direct the gonads to become testes, which produce male hormones such as testosterone. Testosterone cues the development of internal and external male genitalia, as well as the cells that will later produce sperm. In the absence of a Y chromosome, the gonads develop into ovaries and the fetus becomes a girl.

Making a normal male baby requires that the development of Sertoli cells be undisturbed. Anything that interferes at this critical stage of embryonic life can affect whether the baby survives to birth, what his sex organs will look like at birth and throughout life, and eventually his ability to have children of his own. A tragic demonstration of the importance of this window of development comes from the offspring of women treated with diethylstilbestrol (DES), a drug mistakenly thought to prevent miscarriage, between 1940 and 1971. Daughters and sons of women who were treated with DES—sometimes only for a single week during the critical stage of development—have genital abnormalities and infertility problems.

Whether the decline in births of baby boys is connected with these changes in male reproductive health has also been the subject of intense controversy among public health scientists for several years. Several studies—some from the 1980s and early 1990s and others from 2002—show that since roughly the 1970s, the boy-to-girl birth ratio has declined in England and Wales,[51] Denmark, Sweden, Finland, the Netherlands, Germany, Chile, Argentina, Brazil, Bolivia, Peru, Paraguay, Ecuador, Venezuela, Colombia, and Costa Rica. Since 1970, the proportion of boys born in the United States has dropped from 106.568 boys per 100 girls to 105.212; Japan's dropped from 105.888 to 105.004. Even such small drops in the sex ratio can amount to thousands of missing baby boys. For example, if the sex ratio in the United States had remained at its 1970 rate, almost 100,000 more baby boys would have been born between 1970 and 2000. No one knows why this is happening, but some interesting clues exist.

Sometimes conditions arise that distort hormones during pregnancy, or affect sperm shape, strength, speed, or structure. These distortions can throw off kilter the exquisitely sensitive chain of events that lead to normal reproduction and development. A lower level of androgens than is usual may be more common in older fathers or those who work in certain industries. These lower levels of male hormones inhibit the functioning of the boy-making cells, preventing the development of normal male traits. Such a baby will have no testes but will have a tiny penis bud that looks somewhat lumpier than a normal clitoris. These babies are invariably recorded as female at birth, but later in life they will not be able to reproduce as normal females. Except for Olympic athletes, who may undergo genetic testing, many never find out that they are genetically male.

The difference between a male and female all comes down to that Y chromosome. Males, with their XY chromosomes, lack the extra piece of the X leg to stand on. The Y chromosome appears to convey some sort of genetic vulnerability that the X does not have.

Many diseases hit males harder throughout life. In the small Japanese city of Minamata, heavy mercury pollution during the 1950s damaged the nervous system of many residents and caused crippling birth defects and mental retardation in babies exposed in the womb. More boys than girls were born with fetal Minamata disease (FMD), as these defects came to be called. In the times and areas of heaviest exposure, fewer boys were born. The reason may be that males are more susceptible than females to damage from contaminants like mercury. When exposures were heaviest, the males that would have been born with FMD were so severely affected that they did not survive.[52] Similarly, studies of children exposed to lead have found that boys suffer greater loss of IQ than do girls exposed to the same levels; in males lead has also been linked to criminality[53] and reduced fertility.

Today, two out of every three stillborn Japanese babies are male, and fewer baby boys are being born, although no one has been able to explain why. Could it be that the changes in the ratio of boys to girls born in Seveso, Granada, Minnesota, and Minamata are part of a syndrome that arises from the environments of parents? Whatever the causes, it is extremely unlikely that such parallel trends would arise in so many different countries just by chance. The very breadth of the effect points to an environmental cause. Could this widespread decline in male births stem from undetected shifts in exposures to environmental toxicants or other environmental factors? Scientists have speculated that contaminants that change hormone levels during pregnancy may be inhibiting the development of male traits or that other environmental toxins that affect fathers before conception may be disproportionately affecting male fetuses, so that fewer males survive to birth. Still other researchers maintain that the widespread drop in male births is simply part of the natural rise and fall of the balance of the sexes and is not cause for concern.

Unless current research priorities are fundamentally changed in a big way and soon, we will never get the answer to these questions. Some industry public relations efforts have been so effective at creating the impression that there is no problem, that little momentum has been generated to address these issues in the serious manner that they require.

When it comes to understanding what could lie behind these drops in births of baby boys, there are compelling biological reasons for believing that whatever factors are altering the appearance or phenotypic expression of sex at birth could also be tied with many other increasing defects in male reproductive health. The biology on this point is clear. Processes that take place early in pregnancy can affect whether a pregnancy will end, as well as the capacity of any baby to become a healthy young adult who can reproduce when and if he chooses to do so. Unfortunately, the topic has not been accorded the sorts of funding it merits. A series of efforts to discredit and debunk those who raise concerns about children's health issues have effectively stifled the sorts of work that could clarify the problem.

In a report to the Chlorine Chemistry Council dated September 7, 1994, Jack Mongoven, a public relations expert hired to monitor the activities of environmental groups, warned that the issue of children's health, reproduction, and the environment could become explosive:

> Anti-chlorine activists are using children and their need for protection to compel stricter regulation of toxic substances. This tactic is very effective because children based appeals touch the public's protective nature for a vulnerable group and that makes it difficult to refute appeals based on its needs. This tactic also is effective in appealing to an additional segment of

> the public which has yet to be activated in the debate, particularly parents. By characterizing children as the biggest losers of [sic] toxic exposure, the activists have secured an approach that will attract more mainstream support for their anti-chemical, anti-chlorine agendas.

It's important to understand the reasoning here: In Mongoven's view (or at least in the view he's paid to express), a group of people, the "anti-chlorine activists," have conceived an irrational hatred of compounds made from this chemical. They use scare stories about the health of children as a device to recruit others to their bizarre beliefs. Needless to say, Mongoven has it exactly backward. Many of us who are not moved to have personal feelings about particular species of atoms or even molecules nonetheless care passionately about our children's health. We want to know about, and thus defend against, threats to our children.

To deal with the charges of environmental health threats, industry created a potent resource. The Advancement of Sound Science Coalition (TASSC) was formed in 1993 to advocate "sound science in public policy decision making." In outward appearance, the coalition was a grassroots organization of scientists and policy makers concerned with maintaining high scientific standards. Who could be opposed to that? Unknown to many who joined it, TASSC was funded by Phillip Morris, which put up $320,000 for the first half of the year, and was operated by public relations firms; its true purpose was to cast doubt on human studies of environmental hazards, including passive exposure to tobacco smoke.[54] Although the group no longer exists, its Web site, www.junkscience.com, was still going strong in 2002. Here is how the site defines the term *junk science*: "Junk science is bad science used by: personal injury lawyers to shake down deep pocket businesses; the 'food police' and environmental Chicken Littles to fuel wacky social agendas; power-drunk regulators; cutthroat businesses to attack competitors; and slick politicians and overly ambitious scientists to gain personal fame and fortune."

The public relations experts who set up TASSC understood that any group directly funded by tobacco that challenged how federal agencies used epidemiologic studies would not pass the laugh test. They reached out to the Chemical Manufacturers Association and others producing food, plastics, chemicals, and packaging as natural allies. What exactly was their intent? Prompted by EPA's decision to list secondhand or environmental tobacco smoke as a proven human carcinogen, Phillip Morris's vice president for corporate affairs, Ellen Merlo, wrote to the company's chairman, William Campbell, "our overriding objective is to discredit the EPA report and get the EPA to adopt a standard for risk assessment for all products."

All this effort was directed at what appeared to be a narrow technical matter. Studies on passive smoking consistently found that among nonsmokers who had lived with smokers their entire lifetimes, the risk of dying from lung cancer was increased by about 30 percent, or some three thousand lung cancer deaths in a year. This is expressed as a relative risk of 1.3; it means that those who are exposed to passive or sidestream cigarette smoking have 30 percent greater chance of developing lung cancer than those who do not live with smokers.[55]

The industry sponsors had a simple agenda. They would argue that only those increases in risks greater than 2.0, in other words, a doubled relative risk, are important for public health. Risks of less than 2.0 would be labeled "weak." Given that most environmental hazards are distributed broadly, if not evenly, in societies, most increases in relative risk that have been detected in epidemiologic studies of the environment are well under twofold. With this recasting of risk, the great majority of the connections between environmental factors and health problems would automatically be deemed not worth worrying about. There was no precedent in epidemiology or biostatistics for this label. In fact, where common exposures such as air and water pollutants are at issue in a population of a quarter of a billion people, a relative risk of 1.3 or lower can involve thousands of cases. Nevertheless, TASSC made remarkable headway on this position, holding what appeared to be independent seminars with titles like "Good Epidemiologic Practices in the United States, U.K., European Union, and China." At each of these meetings relative risks of less than 2.0 were soon referred to as "weak."

Stan Glantz, professor of medicine at the University of California San Francisco, has received awards from the American Lung Association for his efforts to uncover the secret history of the tobacco companies.[56] With his colleague Elisa Ong, he reported in the *American Journal of*

Public Health in 2000 who had masterminded this activity.[57] James Tozzi, then the so-called chairman of federal focus, was under contract to Phillip Morris in 1993 for $40,000 a month and for up to $600,000 in 1994 with orders to intensify the debate on how science uses epidemiologic evidence, especially any effort to combine studies and assess human risks. This is the same James Tozzi who pulled the plug on early efforts to document the link between air pollution and human health, when he worked as a deputy director in Nixon's Office of Management and Budget during the early 1970s.

By the end of the 1990s, the effort to get epidemiologists to accept the industry definition of weak relative risks had begun to flounder. In 1999 and again in May 2000, the Toxicology Forum, a group with extensive membership from industry that seeks to present itself as an independent research organization, included discussions on so-called good epidemiology practice. These forums included the advice of tobacco industry consultants. But epidemiologists started to balk. Having made no headway with the scientists, TASSC dissolved itself in all but name, leaving its Junk Science home page in the hands of the journalist and lawyer Steve Milloy, for whom the controversy over male defects is merely another opportunity for major media assaults.

In an article published in the *Journal of the American Medical Association (JAMA)* in 1996, some colleagues and I reported declining trends in male sex ratios in the United States and Canada. We asked whether the cause of this decline might not also account for increasing birth defects of the penis and testicles, increasing testicular cancer, and declining quality and quantity of sperm.[58] A similar set of issues has been raised by scientists from Denmark.[59]

For reasons that will shortly become obvious, let me remark in passing that *JAMA* is one of the most prestigious medical journals in the United States. Most submissions are rejected, and those that are accepted must first pass rigorous peer review.

Our article began by reviewing the state of research on the subject. At this point, we noted, there is no debate that distortions in sex ratios have occurred in some groups of highly exposed workers. Several studies found that when highly exposed, DBCP workers fathered more girls than boys. In addition, pesticide-exposed men in the Netherlands were found to father one-half the normal proportion of male children—0.248. Five studies of polluted residential areas in Scotland located near metal smelters, steel foundries, and incinerators also found significantly diminished sex ratios.

We noted that a number of things are known to influence the male proportion of live births—age differences between the parents, older age of father, mother under stress, multiple sclerosis, less-frequent intercourse, and test-tube fertilizations. But although these factors may be relevant to some populations within a given country, none is shared so widely by all the relevant countries that it could account for the steady declines in the births of boys.

Whatever role the environment may play in producing these defects, we said, it cannot account for all these unexplained patterns in so many countries. We know, for instance, that stress elevates some hormones, such as corticotropin. Elevated levels of corticotropin in men lower testosterone and increase estrogen, which, in turn, leads to the production of more female offspring. Women with increased corticotropin levels will have higher levels of testosterone and a hormonal milieu that yields more male offspring. Giving males testosterone before conception increases the proportion of male offspring in humans as well as in experimental animals. In one study, men who were given methyltestosterone therapy fathered forty-five boys and seventeen girls. The proportion of baby boys drops with increasing age of the parents. Still, there is no way that epidemics of stress throughout the industrial world, or simultaneous changes in the ages of parents, can explain the persisting declines in the births of baby boys.

We argued that the reduced sex ratio should be regarded as "a sentinel health event that may be linked to environmental factors," in other words, a canary in the mine of reproductive health. In any event, sex ratio is not static. The boy-to-girl ratio increased in many countries between 1900 and 1950 as better prenatal care reduced the number of stillbirths, which tend to affect males disproportionately. Thus, modern medicine has been able to keep more baby boys alive. But sometime between 1950 and 1970, the male proportion of live births began to decline again. Henrik Moller of the Danish National Research Foundation looked solely at men in Denmark, Finland, Norway, and Sweden.[60]

He argues that there could be a biologic connection between declines in the male proportion of live births in these countries, increases in testicular cancer, and reduced sperm quality and quantity. He suggests that all these effects may be caused by prenatal exposures to chemicals that act like dioxin and the pesticide DBCP, or other environmental agents.

If you understand nothing else from this account, it should at least be clear that my colleagues and I are not mavericks on this subject. We reported on studies by a number of researchers from respected institutions around the world and published in top scientific journals, and didn't take issue with any of them. Despite its disturbing implications, in purely scientific terms the paper was not especially controversial. It did, however, pull these studies together to suggest a larger pattern. Here is what the Junk Science home page said about our publication:

> The amazing and incredible Devra Lee Davis is at it again. . . . Davis now claims that the proportion of males born has declined and that this can "be viewed as a sentinel health event that may be linked to [manmade chemicals in the environment, so-called "endocrine disrupters" or "environmental estrogens"].
>
> But Davis' claim is not based on a scientific study. This article is merely a collection of anecdotes woven together to cause alarm. For example, Davis points to the low male birth rate among parents with the highest serum levels of dioxin from the 1976 industrial accident in Seveso, Italy.
>
> While true, Davis fails to balance that with, for example, the higher-than-expected male birth rate among the most-exposed parents in the Ranchhand-Agent Orange cohort.
>
> But the tale is really told in Davis' allegation that between 1970 and 1990, an "extra" 38,000 female births occurred out of roughly 70 million live births. This equates to about 1900 "extra" female births annually out of the 3.1 million to 4.0 million live births that occurred annually between 1970 and 1990. On a statewide basis, that's an "extra" 38 girl births per state per year or about 3 "extra" girl births per state per month.
>
> How can she identify 3 "extra" female births per state per month AND THEN link the "extra" female births to manmade chemicals? Only Devra Lee Davis knows. You may want to ask her yourself.[61]

It's not clear to me how (to take but one example) Moller's statistically rigorous study of the entire population of Scandinavia born over two decades constitutes an "anecdote," but let it pass. As a lawyer, Milloy acknowledged that what we wrote about Seveso, where exposures to dioxin were high and validated, was true. He then went on to allege that another study, conducted in Vietnam, in which the exposures were not measured, failed to find this same effect. He provided no reference to support his claim. In his world, references are apparently less persuasive than proof by insinuation. He then made up an artificial division of how these extra female births would be distributed equally among all U.S. states, and accused me of making a mountain out of a molehill. His arithmetic is accurate and could have been continued: I calculate a rate of 0.000002 "extra" female babies every 1.6 seconds. It is also sublimely irrelevant.

As an example of how shoddy the Junk Science critique can be, the page included in the spring of 2002 a link to a National Academy of Sciences (NAS) study, "Hormonally Active Agents in the Environment," with claims that this report finds no evidence of human harm.

Here is what the report actually states:

> With respect to the end point most closely studied, sperm concentration, retrospective analyses of trends over the past half-century remain controversial. When the data from large regions are combined and analyzed, some data sets indicate a statistically significant trend consistent with declining sperm concentrations. However, aggregation of data over larger geographic regions might not be an appropriate spatial scale for this analysis, given the significant geographic heterogeneity. The current data are inadequate to assess the possibility of trends within more appropriately defined small regions. Acquiring data at smaller regional scales is critical to assessing the significant geographic variation in sperm concentration.
>
> Laboratory studies using male and female rats, mice, and guinea pigs and female rhesus monkeys have shown that exposure of these animals during development to a variety of concentrations of certain HAAs (e.g., DDT, methoxychlor, PCBs, dioxin, bisphenol A, octylphenol, butyl benzyl phthalate [BBP], dibutyl phthalate [DBP], chlordecone, and vinclozolin) can produce structural and functional abnormalities of the reproductive tract.

This is how science talks. What the NAS committee was saying is that animal studies suggest there's a problem, and some epidemiologic studies suggest there's a problem, but the latter are not sufficiently fine-grained to show an effect on the small geographic scales on which the problem, if it exists, is likely to be found. In other words, there's cause for concern but the right studies haven't been done. This is, of course, a long way from a finding of "no evidence of human harm." In fact, the NAS explicitly noted evidence of human impairment, contrary to Milloy's avowals.

In trying to understand whether there are patterns in sperm counts over time, scientists face a quandary: Because no data have been gathered for this purpose, we are forced to rely on information obtained for other reasons. Typically, this means using sperm counts of donors to fertility clinics, who are often underemployed young men or medical students—not typical of the population. Or it can involve looking at sperm counts taken from patients at fertility clinics or from men who have decided to have vasectomies. These samples are obviously not fully reflective of the U.S. population either. The lack of representative samples is one reason the NAS says more detailed studies are needed.

Here is what the NAS recommended, based on its review of the literature:

> Wildlife and human populations should continue to be monitored for abnormal development and reproduction. Studies of wildlife species that exhibit population declines, abnormal sociosexual behavior, or deformities should be designed to investigate those phenomena with regard to chemical contamination. In human populations suspected of being affected by HAAs, prospective and cross-sectional studies using cohorts tracked from conception through adulthood are particularly needed on female and male reproductive end points, such as sperm concentration, cryptorchidism, and hypospadias. Regional differences in those end points should be studied prospectively to determine whether the differences can be associated with genetic and environmental factors.[62]

So the NAS report and dozens of other investigations agree that animals are in trouble, but for humans we need more studies. Since 1997, the Danish, British, and German governments have all issued reports on troubling trends in both human and animal male reproductive health in the past five years. In 2000, Alex Kirby found that among seven hundred eighteen- to twenty-year-old Danish army recruits, half showed sperm counts low enough to make it hard for them to father children.[63] Still, some argue that we cannot be sure whether these trends are real, or what may lie behind them.

Ecologists have studied reproductive disorders in the beluga whales of the Saint Lawrence River estuary in Canada, the polar bears of the Arctic, and roadkill in the Bitterroot Forest of Montana. The more fat in an animal, and the longer it lives, the more contaminants it can store.[64] And because contaminants of all kinds tend to grow more concentrated as one moves up the food chain, any health problems they cause are most likely to show up first in top predators.

Whales and polar bears are pretty indiscriminate feeders. Beluga whales use their huge jaws to shovel mussels in by scooping them up along with the sediment in which they sit. Blue mussels in the Saguenay River of the Saint Lawrence region have two hundred times more hydrocarbon contaminants in their tissue than do mussels from other regions; these contaminants are absorbed by the whales that eat them. Polar bears eat seals and walruses, when they can find them, which eat smaller fish, which eat tiny animals, which eat microscopic phytoplankton. With each step up the food chain, toxins grow more concentrated. Fat tends to concentrate some toxins quite readily.

Normally, wild animals rarely if ever get cancer, and the rate at which they have ambiguous genitals is also believed to be extremely rare. The rate of cancer in the Saint Lawrence beluga whales is higher than in any other population of wildlife ever studied, occurring in nearly one in five carcasses autopsied since 1985.[65] The levels of persistent PCBs and other organochlorines found in these whales would qualify them as toxic waste. In 1989, a beluga whale from the Saint Lawrence was found with two ovaries and two testes—believed at that time to be only the fourth hermaphrodite ever found in the wild. In 1998, scientists on Svalbard, an Arctic Island off Finland, reported that more than one out of every hundred polar bears was hermaphroditic.[66] The bears are not alone.

Ted Kerstetter is a zoologist, just retired from Humboldt State University on the north coast of California. He now works with the Friends of the Bitterroot Valley, between Hamilton and Stevenville, Montana, tracking some bizarre patterns. Chinook salmon of the last free-flowing stretch of the Columbia River of the Pacific Northwest and fish in other regions are showing up with both sex organs. "Three out of every four fish look female and the males are just not genetically normal. It ain't very pretty," Kerstetter noted.

Judy Hoy grew up on a farm, milking cows day and night, and has the arm muscles to show for it. She runs a licensed wildlife rehabilitation facility in the Bitterroot, using roadkill to feed owls, eagles, coyotes, abandoned baby bears, and other rescued carnivores. In 1995, Hoy began to keep records of what looked like increasing numbers of deformities showing up in the genitals of dead male deer.

In 1999, Kerstetter began working with Hoy and confirmed that the rates of damaged reproductive organs in these deer, including twisted scrotum, undescended testes and microscopic phalli, were higher than any he had ever seen in his more than forty years of working with wildlife. Moreover, the incidence appeared to be growing. Neither Kerstetter nor Hoy received any funding for this research; they did it as a public service. Their work was published in 2002 in the *Journal of Environmental Biology.*[67] Nobody knows what is causing these deformities, but aerial spraying for weeds in the region has also increased during this time. No official study is under way, nor is any contemplated by state authorities.

All this sexual confusion in wildlife probably is not good for us either. How can we be sure, in a world that's constantly changing, what it may mean? How do we know what normal private parts should look like in roadkill? Perhaps we are just getting better at finding strange things. But the alternative explanation, that something more serious is going on, cannot be rejected. We are surely different from deer and whales and polar bears, but we may not be different enough.

Part Three

THE VIEW FROM OUTSIDE

In the Clinton administration, I held a number of different posts. Unfortunately, every time my status within the government grew, my power to bring about real change shrank. My last assignment—to which I was appointed by the president and confirmed by the Senate in 1994—was to help run a new, independent executive branch agency, the National Chemical Safety and Hazard Investigation Board. It was supposed to be a sort of National Transportation Safety Research Board for accidents in chemical plants. For nearly three years, the administration and Congress never approved funding for it, but to me it hardly mattered. By that time I was so far inside, I could not do or say a thing to criticize what was going on.

One of the first signs I ever had that government service was not going to be a long stint for me came when Vincent Garry, a professor of pathology at the University of Minnesota whose innovative work I had cited in a recent article, introduced himself at a meeting at the National Institutes of Health.

"Hello, Dr. Davis," he said, in a somewhat more familiar tone than I expected from someone I had just met. "I've heard all about you. You just got an award from the National Cancer Institute. Congratulations." He looked like the Cheshire cat.

I looked at him askance. "What have you heard?" I asked.

"Well, let's just say that the boys don't expect you to be around long." He shrugged.

8

EARTHQUAKES AND SPOUTING BOWLS

Our people would rather drive than breathe.
—ALEJANDRO ENCINAS

AROUND MIDDAY ON September 7, 1999, I was trapped in total darkness in a stalled elevator in Athens, Greece. I quickly realized that this was no ordinary pause, as the shaft lurched sideways, screeching and groaning like a derailed train. Parnassus and Delphi had obviously survived plenty of earthquakes. I had no idea whether the modern hotel in which I was trapped had been around long enough to have survived a single tremor.

In what felt like the longest minute of my life, I pried the elevator doors open and found myself facing a two-foot-high wall. Climbing out of the opening onto a black-and-white marble floor that had begun to ripple, I joined a crowd racing from the hotel and ended up on a grassy knoll overlooking the sea a few hundred feet away. We would later find out that the thirty-second quake—the strongest in a century—had left more than 140 dead and 100,000 homeless and had caused more than half a billion dollars' worth of damage.[1]

Most of the people around me had come to Athens for the same reason I had: to attend an international conference on public health. In this situation our scientific credentials did us no good at all. None of us knew a thing about earthquakes, except for one woman, who told anybody who would listen that, from her experience, things could suddenly become much worse. Nobody wanted to hear this. But we instinctively moved far enough away so that the massive building we had just fled would not fall on top of us. Within an hour, having succumbed to the universal desire to get back to normal, we headed back inside to hear the final conference papers.

That night, as aftershocks rumbled on, my only concession to this brush with death was to sleep with shoes on and a headlamp and water bottle by my side. I left the large glass door to my balcony open, in case I needed to throw myself off. I later learned I wouldn't have had a chance if another big quake had hit. Now that I've survived the experience, this is one of many things I'm glad I didn't know at the time. The woman who tried to warn us was correct when she said that aftershocks can be even more deadly than the first tremors. Next time I'll sleep outside.

An earthquake, especially a big one (this one measured 5.9 on the Richter scale), seems like a discrete event. We certainly remember it that way: There is the time before, the unimaginable during, and then the relieved or despairing aftermath. But in another sense an earthquake is part of a continuum. The earth is constantly trembling, jiggling, and groaning, in movements that range from massive upheavals down to the rumblings of passing trucks. These smaller tremors are more frequent by many orders of magnitude. Just as geologists understand the big quakes by studying the smaller ones, we in public health understand the Donoras and Londons of our world by studying what happens every day. In public health, those brief, massive shocks are far outweighed by the accumulated effect of the ordinary.

Sometimes the aftershocks never come.

One of those last papers at the Athens conference was mine. I spoke about growing up in Donora, a town where death came from breathing some still unknown fumes from the local mills and coal stoves, but where people wanted to believe it came from an act of nature much like the earthquake we had just gone through. I explained that when I was a girl, nobody ever spoke about the fact that people had suddenly died in my hometown. No studies had been made of the long-term survivors or those who had lived through similar episodes elsewhere. No one had ever demonstrated what caused nearly half the town to become ill and so many to die so quickly. I told my listeners that most of the records gathered on the Donora smog had burned in a mysterious fire years later.

At that point, so soon after the earthquake, the atmosphere was one of low-key, hair-trigger anxiety. I felt like someone who had suddenly learned she had a secret, automatic defibrillator implanted in her chest and had no idea when the next jolt would come. The seriousness of the quake would not register until scenes of rescues and devastation just a few miles away began appearing on the hotel's lobby televisions. As I finished my talk, the clank of a heavy metal chair being pulled across the marble floor sent half the room fleeing to the exits, and the other half dropping under tables. We laughed because we could. Only two miles away, buildings continued to collapse.

Few of the world experts on the environment and health at this meeting had heard of Donora. Right after my talk, a handsome young Mexican doctor came up to me and said, "You have an important story, Dr. Davis. I hope you will be able to tell it in Mexico. We need to hear it. People in my country have never heard about Donora."

I had never met this man, but I knew him by reputation. Victor Borja-Aburto had been a top student at the University of North Carolina School of Public Health. Already armed with an M.D., he had done award-winning research for his Ph.D. dissertation and then, with opportunities at the most prestigious universities in the United States and Europe, he chose instead to return to Mexico, to the National Institute of Public Health. There, working with Carlos Santos-Burgoa and Isabelle Romieu, who also had returned to Mexico with both medical and doctoral degrees, he had continued efforts to quantify what other scientists had observed, more crudely, years earlier.

The importance of the three scientists' work reaches far beyond Mexico. As the world's largest city and the one with the greatest concentration of children, Mexico City offers a statistician's delight. The law of large numbers says that the larger the sample you are studying, the easier it is to find a statistically significant (i.e., nonrandom) pattern. The sheer size of Mexico City's population and the regular monitoring of its air pollution by a system more sophisticated than that of most U.S. cities means that researchers are able to document small changes in health that would not be detectable in smaller cities. Mexico City also had some of the dirtiest air in the world.[2]

For the past decade, Borja-Aburto, Santos-Burgoa, Romieu, and a number of other skilled analysts have devoted themselves to documenting the persisting health threats in Mexico and working to see that policies are implemented to change these risks. Once Romieu made clear that lead threatened the brains and futures of Mexico's children, the country began removing this heavy metal from gasoline in 1986. Within three years, concentrations in air fell nearly tenfold, and that in the blood of schoolchildren in the cities dropped by half.[3] Other sources of lead, such as pottery and burning batteries, would prove more difficult to address, leaving many of Mexico's children with dangerously high lead exposure.

In making the case for public health actions, Mexico's public health leaders face serious challenges. First, they have to establish the true rates of certain diseases in populations of various sizes and compositions. Second, they are trying to connect patterns in pollution with those in health, taking into account the fact that health is heavily affected by poverty and malnutrition. It is not a job for the faint of heart: Every day, they can see and smell the reasons their work is so important. The city has more than 20 million residents and more than 35,000 different industrial facilities, all releasing various pollutants into a valley that sits some 2,200 meters above sea level and is surrounded by even higher mountains.

While some of their questions are decades old, the three scientists are supplying answers of unprecedented precision. In the early 1990s they showed that with every increase of just 10 micrograms—ten millionths of a gram—of fine particulate pollution per cubic meter of air, the risk of death for the population rises 5 percent. How can this be?

To come to this conclusion they have relied on and refined the statistical modeling that Lester Lave and Eugene Seskin introduced in the early 1970s. Their work takes into account smoking, family wealth, temperature changes, long-term patterns of many different types of pollutants, and various other phenomena. It also uses an averaging concept that looks at rates of death or disease relative to levels of air pollution over the previous few days. In Mexico, hourly levels of the gaseous pollutant ozone exceed the national standard of 110 parts for every one billion molecules of other gases (ppb) most of the time.[4] In Mexico City, for each four-day period that the hourly average of ground level ozone rose 100 ppb, the death rate increased between 2.4 percent and 3.9 percent for those over age sixty-five. With every increase of 10 micrograms per cubic meter of total particulates over a four-day period, the rate at which all people die rises 6 percent.[5] In a city as large as Mexico these numbers translate into more than 6,400 deaths each year.[6]

These numbers create a problem. When looked at all together they are so large and so impersonal, it is easy to forget that behind them lie real people, each of whom dies alone. Their deaths happen as most do, in hospitals, sometimes in clinics or in homes. Seldom do they occur in clear ways that signal that air pollution has caused the deaths. Because so many younger persons will die and each one of these may lose more than sixty-five years of life, work, and love, statisticians sometimes estimate what is called total years of life lost. By this calculation, if one in ten deaths in Mexico occurs in an infant, then each year, Mexico loses the equivalent of 41,600 person-years of life from air pollution alone. The heartache behind these numbers cannot be calculated.

Similar patterns have been uncovered in cities throughout the world, with children suffering the greatest harm no matter where they live. Cultures, places, and times may differ, but the way the lungs of the very young and the elderly respond to fine particles and ozone appears to be fairly constant.

Borja-Aburto had turned down offers from U.S. and European institutions because he wanted more than a safe academic perch from which to add to the published literature on proven hazards. With many nieces and nephews living in Mexico as well as two children of their own, he and his physician wife have a deeply personal stake in the matter. He had heard that shortly after the Athens conference, I would travel to Mexico as part of a public conference to bring attention to the city's growing problems with air pollution. I was scheduled to present the results of a simple exercise conducted by researchers from the World Health Organization (WHO) and the World Resources Institute (WRI), in which we had ranked and compared where the most children breathed the most dirty air. Working with experts from several countries, we had used data from WHO's Global Environmental Monitoring System from more than two hundred of the world's largest cities for the years 1993 to 1995 to show that the health effects of pollution were a far more massive problem than anyone had realized.[7] Annual average levels of air pollutants in many cities ran between two and eight times the WHO recommended levels.

Some 85 percent of all children live in developing countries, and half of those live in cities. But children are not just miniature adults: Their lungs, hearts, immune systems, reproductive systems, and brains are still growing. If on top of malnourishment and inadequate health care, they are taking in pollutants with each breath, their ability to mature, learn and ultimately to work may all be impaired. Even if they move to a clean environment later on, damage to their brains or lungs may never be repaired.

Borja-Aburto knew all about these difficulties firsthand, along with the benefits of public support. He and his seven siblings had grown up in an undeveloped region of south Mexico, in the small town of Guyameo, in the state of Guerrero. His father was murdered by a drunken soldier. His thirty-three-year-old widowed mother worked as a seamstress. All of her children completed college. The family lived in areas so poor and crime-ridden that neighborhood drifters each morning would make sure they caught the school bus safely. Moving to the city—even with its crime and dirt—for his publicly funded university training had made it possible for him to end up at the University of North Carolina. He understood the need to make cities safer and cleaner in a way that few others could.

The work I was scheduled to discuss in Mexico had already been presented to the annual meeting of the American Association for the Advancement of Science (AAAS) in January 1999 in Anaheim, California. AAAS invited us to provide a press briefing identifying the largest cities in the world where the most children regularly breathed the worst air. The exposure of so many of the world's children to heavily polluted air was then a new public health issue. Our findings, which had been posted on a number of Web sites and released as a WRI publication, were picked up by National Public Radio (U.S.), the British Broadcasting Corporation, *Asahi Shimbun, China Daily, South China Morning Post,* the *Guardian* (London), and the *New York Times Week in Review.*

At the top of our list of polluted urban areas was Mexico City, with its 20 million residents, half of them under fifteen. The *Times* reported that the drawings of local children usually featured brown skies. Pollution is so heavy that the children had little experience with blue horizons. We were scheduled to present a summary of our work to a conference on air pollution organized by the Climate Institute of the United States, EPA, the World Bank, and Mexico City's Ministry of the Environment. I had been warned by John Topping, the Climate Institute's president, that our message would not be well received. No country wants to be known as the worst polluted.

Topping's organization had bet its reputation on its ability to get Mexican officials to acknowledge that there were real and immediate actions they could take to improve the air. We inserted a statement into our press release that documented the tremendous and rapid progress Mexico had made with respect to reducing the amounts of lead placed into its gasoline and getting into children's blood and brains. We noted that Mexico had done in four years what had taken the United States and England more than two decades to do. Some officials had argued that since so much of the work on air pollution and health had originally come from the United States, and since each country was unique, we could not be sure whether Mexico faced the same risks. Critics noted that even though pollution levels in Mexico were sometimes more than twice as high as in the United States, health problems were not simply doubled. What was needed, they argued, was a massive research campaign to see whether people outside the United States did not have some sort of different way of responding to pollution. Topping and I did not think that young bodies and lungs in Mexico and the United States were likely to be that different, nor did we think it made sense to wait for more studies before starting to act to reduce the burdens of pollution.

The mathematical nature of the relationship between pollution and health has become one of the central preoccupations of public health research. Just how much does a given unit of pollution increase the litany of health problems, ranging from deaths in infants to hospital admissions for asthma to lost days of work? So many studies now exist showing these relationships in so many different places that they can be combined in what is called a meta-analysis. The prefix *meta* in this context means "overall." Meta-analysis takes all studies done on similar circumstances looking at related illness or death and similar exposures, and pulls them together. This technique has made it quite clear that whether the observations are in Mexico, Shanghai, the Netherlands, or Los Angeles, for every 10-microgram increase in particulate air pollution per cubic meter of air over the previous three days, there is between a 0.6 and 2.4 percent increase in deaths every day.[8] Mexican studies show a larger change in daily death rates than in the United States, perhaps because Mexico's population faces other challenges, such as poorer health care and poorer nutrition. In Mexico, one out of every ten deaths every day occurs in a baby, compared with only one in a hundred in the United States.[9] In both countries, however, air pollution significantly increases the chance that such a death will occur. Reports linking air pollution with the size, weight, and health of babies at birth, and with their chances of living and dying are now coming in from more than fifteen countries.[10] This relationship has been found so often in so many places that it is no longer debated. By the time of our conference in Mexico, we had clear evidence of a dose-response relationship: Children in the dirtiest and poorest areas have the highest rates of deaths, asthma, and other illnesses. But what we did not have in Mexico in September 1999 was an environment in which we were free to speak about these things.

Less than a decade earlier, discussion of public health in Mexico was limited to a small circle of experts. Scientists were permitted to do research but were expected not to make much commotion about their work. One chapter in a book produced by the National Commission on Human Rights in 1992, written by Santos-Burgoa, then the dean of the National Institute of Public Health, and Leonora Rojas, reviewed the extensive literature on hazards linked with polluted air. It concluded that air pollution levels in Mexico were so high, they could be considered a human rights violation. Seven months after the book appeared, a journalist paged through this compendium and created a front-page story for the *Excelsior,* headlined "Air Pollution Kills Mexicans from Cancer."

In response, Mexico City officials from many different departments convened a press conference to explain the situation in some vague way, as though their mere appearance would itself assure people that things were not that bad. Santos-Burgoa, like Borja-Aburto, was encouraged to keep quiet but was invited to attend. The officials explained all the activities under way but did not address the question of how dangerous the city's air pollution levels were in the first place. They were so evasive on this subject that at one point a journalist asked the authorities, "Are you contradicting what Dr. Santos-Burgoa has written?"

The reply was straightforward. "No, of course not. You can ask him yourself to explain what he meant."

The entire room turned around to look at Santos-Burgoa, who stood out, at six feet two, in the back of the room. Even though he was dean at the tender age of thirty-nine, Santos-Burgoa had little experience with the rough-and-tumble of politics and was stunned to find himself the center of attention. He was asked whether he now wished to repudiate what he had written.

"No. I defend every word in the book," he replied.

With this answer he became known as someone the ministry could not trust. Within a few years of this incident, he left his post to set up a nongovernmental organization.

When I was planning my trip to Mexico in 1999, I was ignorant about this history of subtle intimidation. I asked Borja-Aburto why he was concerned about my plans to discuss the growing literature that he and others had created.

"Look, I'm glad you're coming," Borja-Aburto told me when we spoke in Athens. "We need you. But you cannot mention my work or me by name. I can do whatever research I want, but I am not allowed to discuss my findings in Mexico." He was calm and clear.

"What's the problem?" I asked. "You've published this work in a major journal with one of the leading experts in the field. Other papers have come out saying the same thing. You're a physician. You were a top student at your university. What's going on?"

He offered a puzzling reply. "Yes, my papers have been published, but they are just not talked about here. I am allowed to write in academic journals. But if you start mentioning my work in Mexico and referring to me personally, it would not be good. Believe me."

I was flabbergasted. But I knew that I had to be careful. I was a scientific tourist. He had to live with the consequences of my visit.

On September 19, the eve of the conference, the anxiety was palpable. A barrage of e-mails and phone calls, many from folks I had never even met, warned that Mexico was in the midst of upheaval on so many fronts that this issue of pollution could become explosive. Three major political parties were involved in the first truly competitive election for president in living memory. The oxymoronically named Institutionalized Revolutionary Party, or PRI, had not lost an election for seventy-one years. Its candidate, the bureaucrat Francisco Labastida, gave new meaning to the word lackluster. In a sign that some political change was possible, the left-of-center Party of the Democratic Revolution had recently won the mayoralty of the City of Mexico. But Mayor Cuahtemoc Cardenas, now their candidate for president, was not igniting much enthusiasm.

The charismatic, tall, cowboy-booted and -hatted candidate of the National Action Party (PAN), Vicente Fox, was beginning to look more and more like a serious contender. Fox was making alliances between groups that had rarely even been in the same room before. Just as Richard Nixon had done in the United States in the 1970s, the conservative Fox began to court the environmental vote. He allied himself with the Green Party of Mexico, not because he was necessarily interested in their positions but because he needed the votes. The election mantra had become "Anybody but the PRI."

As the campaign began to get into full swing in the fall of 1999, Mexico was poised for a major shift. After the vote was taken the next spring, the Greens claimed 40 percent. Fox lived up to his name. The Fox candidacy, with its focus on reform, had made it impossible for people to deny the importance of pollution.

Just eight years earlier, on March 17, 1992, the government had ordered children younger than fourteen to stay home from school for a month because of extremely high levels of air pollution. At that time, this did not lead to major public discussion of the issue, in part because of another tradition in Mexico. Politicians had special funds set aside that they could tap to routinely pay journalists not to write stories that they wished to keep secret. Stories about pollution were one subject for which lots of money probably changed hands, although no figures will ever be available on the extent to which this took place. All we really know is that for years the newspapers did not report what most folks regarded as an open secret—Mexico was indeed highly polluted. The front-page coverage given to Santos-Burgoa's analysis of air pollution's damage had been an aberration, and folks would try to keep it that way. It was as though not talking about the problem might somehow make it disappear.

Two weeks after the Athens earthquake, and some eight months before the presidential election, I flew into Mexico. The keynote speaker of the conference I was attending was a modern-day hero of the country—Mario Molina, the brilliant Nobel Prize–winning chemist based at the Massachusetts Institute of Technology. Working with F. Sherwood Rowland at the University of California some thirty years earlier, Molina had made a fundamental discovery about chlorine chemistry, arguing that atoms released from decomposing gases of chlorofluorocarbons (CFCs) could destroy the band of invisible ozone gas in the earth's upper atmosphere.[11] In the early 1980s researchers at the South Pole showed that the chemical process that Molina and Rowland had predicted was in fact depleting the earth's ozone layer. The two were awarded the Nobel Prize for this research in 1995.

Molina is a small man, modest and soft-spoken, with an expansive intellectual reach. As Mexico's first Nobel laureate, he understood that he commanded a rare platform. He turned his scientific prestige and research attention to studying the atmosphere as a public health issue. Molina brought together teams of modelers from around the world, many of them trained at MIT and Harvard, to calculate the full burden on public health and the economy of Mexico City's current patterns of energy and transportation.

It did not take a prize-winning scientist to appreciate the grim reality of air pollution in Mexico City, but it did take someone of that stature to force more public discussion of its realities. In the previous decade, the government had found itself compelled to act on the problem it did not want people to talk about. In 1988 it had begun declaring air pollution alerts and attempted to cut down on driving by ordering cars not to drive on one day a week—*Hoy no circula.* Every car license plate had a sticker showing which of the seven days it could not be driven. This had the effect of boosting the market for old clunkers from the United States—none of which met the emission standards of either country. These old cars were bought by enterprising residents so that they would have enough vehicles to be able to drive one of them, no matter what the restriction. Many were later sold in the black market. The program was an utter failure in every way but one. It showed the ingenuity of Mexico's middle class in subverting any effort to restrict their use of cars.

Molina, because of his stature, was immune to political intimidation. By the time of the election, he had the ear of every presidential candidate. The facts were simple. Mexico City is situated in a high-altitude, pollutant-trapping valley. More than 3.5 million vehicles plus thousands of industries release exhaust into the air each year. Using back-of-the-envelope calculations, Molina's team estimated that a 10 percent drop in pollution in Mexico City would save about 3,000 lives, 10,000 cases of chronic bronchitis, and millions of days of lost work in a single year. This eminent group declared the Mexico City metropolitan area "one of the most polluted regions on the planet." This simple truth had never before been uttered in public without immediately being challenged in vague and subtle ways that actively discouraged frank public discussion. As Santos-Burgoa learned in 1992, there were prices to be paid for public candor.

The message of health scientists with whom I worked only added urgency to Molina's statements. At the September conference we were able to state with some confidence that adopting more efficient energy and transport would have a great social benefit. Because of their young population, great and growing cities like Mexico have tremendous potential to improve their health for the twenty-first century by investing now in more efficient systems of energy and transport.

This was news: For the first time, the international community had come to Mexico to support the need for serious action against pollution in the country. Normally, in giving such a talk at such an event, I would have credited Romieu, Santos-Burgoa, and Borja-Aburto by name, as the key researchers on this issue, especially since I was speaking in their hometown. But not this time. The outspoken Molina had given them cover—we quoted him instead. In almost any other setting, such an omission would have been a stinging insult. Here it was a political necessity.

All the major papers, CNN Spanish, and radio covered the press briefing the day we presented our report. We were joined by the same Mexican environment officials we had been warned would object to what we had to say. The next day the local daily newspaper *Reforma* featured a photo of four grimacing children in cloth masks, surrounded by clouds of exhaust fumes. The front page of *La Prensa*, another large newspaper, read: *Si, Mata el Smog del DF!* (Yes, there are deaths from smog in Mexico City).

As I mentioned, however, sometimes the expected aftershocks never come.

The day after our conference, I was riding in a government car with Mexico City's director general for environmental projects, Diana Ponce Nava. Her cell phone rang. It was an official from the Ministry of Health. I could not follow all the Spanish conversation, but it was clear that we had indeed struck a nerve. An animated discussion was under way. After a few minutes, Diana asked the caller to hold. She handed me the phone.

The health ministry official was calling, he said, because he was concerned about how people might react to the news accounts that smog could kill children. He wanted to issue some sort of a press clarification, not an outright retraction, mind you, just a statement explaining, *clarifying*, what I had said. After all, we did not want people to panic. I explained that I had only reported publicly on work that had been in the published literature for some time. Of course, he agreed, but he felt that something more was needed—not exactly a denial, but some sort of simplifying statement. I asked what he proposed. He said he would work on it and get back to me. I never heard what happened behind the scenes, but this time the clarifying statement never was issued.

One reason the official decided to drop his idea may have been that the popular radio host Manuel Guerra had shared our information with millions. Within a few hours of our conference, both Molina and I had participated in a weekly radio program, *Ecocidio,* which included live (and lively) call-in queries. With support from the U.S.-based Ashoka Society, and despite efforts to suppress his work in the past, Guerra has argued that Mexico City has to stop assuming it can grow indefinitely. He and the thousands of mothers who log on to the Internet every day to learn whether their asthmatic children can go out to play had paved the way for us to relay what we had found.

In retrospect, our conference took place at a watershed time for Mexico. The country stood on the verge of an outbreak of openness and sunshine that it had never before experienced. Within the month of this event, Margarita Castillejos, a collaborator of Borja-Aburto's, had been allowed to do something that would have been unthinkable just a few years earlier. She appeared on Guerra's radio program talking about what parents could do to help protect their children when the air worsened.

Santos-Burgoa, now an official with the Fox administration charged with protecting consumers from a broad array of health threats, noted that since 1999 even bigger changes have taken place. For the first time in Mexican history, the people negotiating environmental issues on behalf of the government understand public health. A trained public health professional, Julio Frenk, heads the Secretariat of Health. Three of the four undersecretaries are also public health professionals, as is a large proportion of the third-level leadership. Most important, the Ministry of the Environment finally has a seat in the Economics Cabinet. As one indication of how far things have come, the government is officially inviting public discussion on the health threats of pollution and what needs to be done about it. March 6, 2002, saw the publication of Mexico's first National Action Plan on Environmental Health. On April 30, 2002, the Mexican Senate unanimously passed a landmark freedom-of-information law, passed earlier by the House of Deputies. This law gives citizens access to information previously kept secret about how their government spends money and makes decisions.[12]

Just as it took a conservative President Nixon to start federal efforts on the environment in the United States, it has taken a conservative Mexican president to convince business, government, and military leaders that the environment is not something that can be fixed later on, but has to be protected now, while development proceeds. People must have enough information to understand how the environment shapes their lives and those of their children and grasp that more money cannot always repair damage to young brains or lungs. They need to hear that building more highways makes more pollution, does nothing to ease congestion, and exposes people driving in cars to much higher levels of pollution than those on the street.[13]

In this different environment, Borja-Aburto, Santos-Burgoa, Romieu, and their colleagues are not only permitted to talk openly about their work, but encouraged to do so. A government strapped for resources cannot possibly address all the issues posed by environmental contamination on its own. It is mobilizing citizens to engage directly in the process so that the children of Mexico get to become adults in a world that is cleaner and healthier than that in which they currently live.

In the summer of 1999, Borja-Aburto, John Topping, and I hiked to the top of the Nevado de Toluca, a 14,000-foot-tall dormant volcano within an hour's drive of Mexico City. We looked out in the direction where we knew one of the world's largest urban zones sat about 7,000 feet below us. As far as we could see was nothing but brown, murky haze. Nowadays big winds or heavy rains are the only things guaranteed to clear up this soiled soup. We vowed to return when we would be able to see the magnificent city we knew lay below.

In China, the release of the same simple WRI study on the world's cities unleashed a similar set of tremors. My colleagues and I knew that the fundamental rule of any large bureaucracy applied doubly in China: Never surprise the boss. In fact, the bosses in China had long known of our work, and we had carried out the project with the cooperation of the Chinese Environmental Protection Agency, but China is a unique country in many respects. Working side by side within the government are hard-line advocates of promoting coal and visionaries who see the need to devise cleaner ways to use existing resources, and to devise resources that produce no pollution at all.

A fierce struggle was then under way within the government as to how to avoid relying on the same technologies that Western societies had employed a half century earlier. With so many people living in such concentrated areas, if China were to go through the same technological history as the United States had, a much greater proportion of its citizens and their environment could be permanently impaired. The resource requirements alone are daunting. If each family in China were to have as many cars with the same types of engines as U.S. citizens do now, the world would need another planet to provide the fuel. The world cannot afford to have its largest country progress through the old, dirty technologies before ending up with cleaner and more efficient ones. Many countries are avoiding installing telephone poles and wires and moving straight to cellular communication. China needs to make a similar technological leap in the area of energy.

In September 1998, as we were developing our WRI report, I had attended a meeting of U.S. and Asian leaders seeking to find common ground on environmental issues. The meeting was hosted by Senator Max Baucus with the Mansfield Center at the University of Montana. After touring the splendor of Glacier National Park, in central Montana, we got down to business. I had with me a first draft of our analysis of the largest cities with the most children at the highest risk of exposure from air pollution, which I showed to the team traveling with China's environment minister. Of the ten worst cities in the world for which we had information on average yearly levels of air pollution, nine were in China. Air pollution for the country as a whole ranked among the worst in the world.

It is often challenging holding meetings with people whose primary language is not your own. But in this case, no translator was needed. I put onto the overhead projector a simple bar chart. It showed that the average yearly levels of air pollution in many of China's cities were four times higher than the levels recommended by WHO. The graph had not been on the screen for ten seconds when the minister himself gestured to me that he would like to take my plastic transparency sheet back to China. I handed it to him on the spot.

China is a land of monumental proportion. The country has a history of taking on enormous projects, which are sometimes undermined by their very scale and scope. In the hills of Nanjing lies one monument to imperial excess—the world's largest almost-tombstone. At the head of a

2-kilometer trail through the forest sits a carved stone 60 meters high, 18 meters wide, and 5 meters thick and weighing 8,000 tons. It was intended as a tombstone to one of China's emperors of hundreds of years ago. More than two thousand workers died carving this stone: They were killed by their overseers if at the end of a day they had not produced enough rock powder to indicate hard work.

In the end, the proposed tombstone proved so massive that it could not make its scheduled slide down an ice river in winter to its intended resting place. The little village of Graveyard now surrounds the ruins of imperial excess, where Chinese lovers occasionally find shelter in its niches. The site of a massive miscalculation, Graveyard markets itself as a remnant of imperial history.

A similarly vast project remains incomplete in the city that our report identified as the worst polluted in the world. Lanzhou, a town of 2 million people, lies deep in the interior of mainland China in Gansu Province, close to the Tibetan Plateau. The average levels of pollution in Lanzhou were more than 100 times the WHO guidelines. The new generation of China in such regions risks permanent damage. Lanzhou faces some special challenges because, like many industrial towns, it sits at the bottom of a narrow, bowl-shaped river valley. The surrounding hills regularly create inversions that trap hot factory and coal stove gases, hazes, dust, and fumes. Starting in 1998, city planners had begun to dynamite and bulldoze the tops off the highest of these hills, thinking that this would allow the pollution to blow away. The effort to make a molehill out of a mountain only threw more dust into the air. The project was stopped after some townspeople whose ancestors lie in tombs in the hills blocked the equipment.

One of the local officials in charge of the process acknowledged the futility of trying to clean up the air using dynamite and bulldozers. "We have achieved a kind of fame for our air pollution," noted Yu Xionghou, the director of the Lanzhou Environmental Protection Bureau. "We're doing the best we can to improve it. . . . I think it was like drilling a hole in the wall of a smoke-filled room. We need to do something else."[14]

Like many cities throughout China, Lanzhou has an ambitious pollution control program, which rests heavily on public outrage for its support and international agencies for funds. Shifting from dirty, brown coal to natural gas; planting more trees; setting up wind-, solar-, and dam-powered electricity sources—all these activities are under way, as is a release of public candor about how bad the pollution has been.

In the winter of 1999, during an air pollution emergency, Lanzhou's environmental agency ordered more than one hundred factories temporarily shut down, including some state-owned refineries and a gigantic steel plant. The factories resisted these orders because this conflicted with the federal government's mandatory growth rate of 8 percent. "It was very, very hard to shut down the factories," said Yu, who described strong resistance not only from the factory managers but also from other government departments intent on meeting their economic targets.

In China, the media, like much of the country, have been highly controlled for generations. But on the issue of the environment, freer media have become a central part of the government's efforts. When the local Lanzhou EPB informed the press that the large factories were directly endangering people's health, the resulting public loss of face forced the factories to do what they had privately resisted, and halt production.

In a land given to excess, the ability to regroup and shift gears becomes greatly valued. This public disclosure of the extent of China's serious air pollution problems became yet another challenge to those determined to keep the country moving ahead. As China prepared its bid for the 2008 Olympics, the last thing it wanted was to win the world prize for air pollution. Recognizing that where there are big problems, there are opportunities for big actions, authorities in Beijing shut down all major polluting industries to celebrate the fiftieth anniversary of the founding of the People's Republic of China in November 1998. Twenty thousand heavily polluting small taxis called *miandis* were rounded up and smelted down to get them off the streets. The skies cleared. We can be sure that people felt better in Beijing then, but we have no data to prove it. One reason we can know that health improved is because of what happened in the United States during the summer Olympics of 1996. In Atlanta, when traffic was curtailed and twenty-four-hour public transportation was made available, both pollutant levels and emergency room admissions for asthma dropped more than 30 percent.[15] One can only imagine what a similar action meant for Beijing, where pollution was higher to start out and the number of people at risk was more than double that of Atlanta. The episode was a kind of anti-Donora, but of course no one studied it.

One of China's unique challenges at the turn of the millennium was that in its official history, there had never been any problems with the environment. In 1970, Mao Zedong had declined to send any delegates to the first World Earth Day. Only decadent Western countries, he explained, contaminated the environment. Under communism, with its heavy rural emphasis and state-run industries, environmental pollution could not occur. Because they recognized no pollution, the Chinese developed no means of dealing with it.

The unleashing of capitalism further fueled the pollution bubble. In 1994, during a visit to the booming southern coastal region of Guanchow, Premier Dong Xiaoping decreed that under communism it was now "glorious to be rich." This led more than 400,000 rural township and village enterprises throughout the country to switch from agriculture to more profitable industrial pursuits. They built paper and pulp mills, metal-plating firms, and other small factories, with no thought whatsoever to their environmental consequence. In a country in which only 15 percent of all human or industrial wastes got any sort of treatment before being dumped into lakes, streams, ditches, or sea coasts, the consequences of this pollution could often be smelled, felt, or seen.[16] In one instance, a proposed chemical plant could not be located on a river in China because the intake water was too dirty.

China had one other distinguishing challenge relating to its children. Official policy permits only one child per family in its urban centers. To enforce this policy, working women routinely had their urine checked for hormone changes that indicate early pregnancy. Only those pregnancies that were approved were allowed to proceed. Each child born in urban centers was very much wanted. Many of the only-children produced by this policy have now grown up, married each other, and had only-children of their own. These only-grandchildren are doted on by six adults, for whom they represent the entire hope for the family's future. It is no surprise that urban Chinese families are obsessed with their children's health and survival. Our studies clearly showed that the health of these children was on the line just from breathing.

Qu Ge Ping, the first head of China's Environmental Protection Agency in 1983, understood early on that China would never progress until it had addressed and fundamentally reversed its polluting history. A small, dapper man with an instinct for survival and a sense of what can be done when, Qu spearheaded efforts to tell the public the full extent of environmental damage in China, even before Mao's death in 1976. By the end of the 1990s, Chinese public television featured weekly series on environmental threats. One documentary showed villagers who had burned down a polluting factory that had damaged their rice fields.

In a country that has never tolerated citizen groups, a host of so-called nongovernmental environmental organizations began to emerge, including China's own Greenpeace and World Wildlife Federation.[17] Once-secret information on air and water pollution began to be talked about in public. This release of previously confidential reports was no accident. Qu and other leaders in China knew that even as environmental protection had been elevated by China's leaders to the same ministerial level as the iron and steel ministry, its staffing had been cut in half. They also had to deal with a bizarre system of funding local environmental protection bureaus. Many of these programs got all their revenue from fines levied on polluters. If pollution were to disappear, in theory, so would any funding for local environmental protection. If China was going to tackle its growing environmental problems, then the public had to become an ardent and active ally in the ministry's efforts to shut down or revamp dangerous polluting businesses and practices.

Within the Chinese bureaucracy, the release of our study on urban air pollution prompted a major tug-of-war between those who chose to embrace and those who ignored our report. We had noted that although Beijing's air pollution was not the worst in China, it was among the worst in all large cities of the world for which we had data. Using its own system for ranking pollution, the Chinese press had begun to publish official records. North China cities during winter ranked among the worst in the world, with half of them exceeding even China's relatively lax standards.[18] Beijing, Tianjin, Gansu, Xinjian, Shaanxi, Shanxi, and cities in the provinces of Henan, Jilin, Qinghai, Ningxia, Inner Mongolia, Shandong, Hebei, and Lianoning—all regularly had levels of pollution several times higher than WHO-recommended levels.

To those who ran China, these results were no surprise. Within a few months, our work had been translated and circulated extensively on the Internet in both English and Chinese. At about the same time, my Carnegie Mellon University colleagues Guodong Sun and Keith Florig issued a study estimating that air pollution in China caused more than 1 million deaths a year, with indoor burning of fuels for cooking and heating having the most severe impacts on women and children.[19] The *South China Morning Post,* based in newly incorporated Hong Kong, ran a story on our 1999 AAAS release with the headline "Nine of the Worlds' Top Ten Worst-Polluted Cities in China." This was one prize for which nobody wanted to claim credit.

Not far from the western slopes of the mountains of Yanshanm on the outskirts of Beijing sits a not very high hill, called Fragrant Mountain. As recently as the early 1980s, the local residents would come to the area seeking fresh air and the shade of its trees. By 1999, however, with the city's population at 13 million, they no longer made the trip for that reason. Pollution from the city often infiltrated here as well. I hiked up the mountain that year, curious to see the view from the top. As with the Mexico volcano, once at the summit, I could see no sign of city life below.

But just as China has a population policy that no other country could ever attempt, it also has the capacity to undertake massive and radical changes that would never succeed elsewhere. In China, those who work for the financially strapped Environmental Protection Agency are pressing the case for massive changes in public transit, stringent reductions in pollution levels, and major transformations in the ordinary way of doing things. Those charged with economic development do not deny the need to act. The latest battles are taking place within the rarefied halls of economic analysis. China's economists have conceded that, in their best estimates each year, the costs of environmental damage from defective ecosystems, ruined water supplies, and polluted air are at least equal to the current rate of economic growth. Clearly, this is not a sustainable condition.

One of the greatest forgotten secrets of history is that possibly half the basic inventions and discoveries on which the modern world rests came from China. Paper, astronomy, fishing reels, manned flight, hot-air balloons, parachutes, natural gas, chess, steelmaking, guns, biological pest control, navigation, underwater mines, pasta, poison gas, printing, clocks—all had origins in China, long before they ever appeared in the West.

Some Chinese inventions have never been duplicated elsewhere. For instance, the strange objects known as spouting bowls were first created 2,500 years ago. Made of finely tuned bronze, the bowls are crafted so precisely that when the right amount of water is placed in them and the handles are rubbed correctly, the bowls generate vibrations that give rise to water spouts. Science historian Joseph Needham reported seeing spouts as high as three feet. Cecil Beaton, the famed British photographer, produced an image of a boy demonstrating a bronze spouting bowl at the Chinese Temple at North Hot Springs sometime in the 1950s.

These spouting bowls are masterpieces of precision casting. They symbolize what China and Mexico are striving to achieve in balancing pollution and economic growth. The water spouts arise because two competing, equal, and precisely opposite waves are generated simultaneously. The waves start from opposing sections of the bowl and move, at the same speed and height, to come together, like two hands beginning to pray, creating a standing wave. First the water in the bowls ripples, then it begins to vibrate. The bowls start to sing, producing a deep resonant tone that becomes stronger as that water stands higher and higher.

By any measure, a standing wave is extraordinary. It precisely balances the dispersive forces that normally would destroy it with other, peculiar, counteracting forces. The combined effect of exquisitely balanced opposition gives the wave the continued cohesion to prolong its existence.[20] A standing wave is thus a profoundly apt model for the ancient Chinese concept of the Tao. All things flow from and must return to the same point of the Tao. Like two opposing waves that will cancel each other out, pollution and growth, if balanced just so, will leave an intact and functioning whole.

9

A GRAND EXPERIMENT

> In the cities with their tall buildings and narrow roads, the pollution that comes from their residents and their waste makes their entire air reeking and thick, although no one is aware of it. And if you have no choice, and you cannot move out of the city, try at least to live upwind. Let your house be tall and the court wide enough to permit the northern wind and the sun to come through, because the sun thins out the pollution of the air.
>
> —Moses Maimonaides

The historian of science Derek Price, in 1963, distinguished between what he called big and little science, though the same rules apply to both.[1] Little science refers to the lone scientist working in isolation to improve our understanding of how the world works. Think of Galileo laboring outside the church's interference, tossing apples and feathers off towers, using the newly developed telescope to develop his radical theories of gravity, the conservation of energy, and the Copernican idea that the earth moved around the sun. Big science requires teams—dozens, sometimes thousands, who work on large projects to accomplish complicated tasks.

There may well be no issue today more complicated and more in need of big science than that of understanding the connections between human activities and our planet's climate. Whether we like it or not, human life constitutes an unprecedented experiment with more variables than can ever be studied at once. We can look with some relief at smoldering Venus and frosty Mars and be grateful that we appear to be the Goldilocks planet—just right, so far. But with respect to solid experimental information on how the lives we lead are affecting our global environment, we are literally driving the bus that we are trying to study.

Understanding global climate is one of the most important challenges the scientific community faces today. Our planet sustains abundant life—ours and others'—because it has a balanced atmosphere that receives the right amount of incoming radiation, holds in enough heat and water, and cycles and releases these and other materials at rates that keep the temperature and chemical composition of the soils, oceans, and atmosphere within a livable range. Whenever anyone burns hydrocarbons—whether dung, wood chips, coal, gasoline, or peat moss—or wherever sheep, cows, pigs, yaks, or chickens are raised, or rice grown, exquisitely complex chemical and thermal reactions ensue. Greenhouse gases like carbon dioxide, methane, and other by-products of natural and human activities tend to warm the atmosphere, which can alter weather patterns, produce more severe extremes of weather, and ultimately shift the world's climate in unpredictable ways.[2]

This chapter describes how two types of gases generated by humans—chlorofluorocarbons (CFCs) and greenhouse gases—threaten to change the nature of life forever. It tells how warnings by scientists that CFCs imperiled the earth's upper ozone layer were at first dismissed. It was a dismissal fueled in part by a well-financed campaign conducted by those who stood to lose the most from changing their production. Global actions to phase out the production of CFCs[3] took root only after scientists had generated irrefutable evidence that damage was under way, and the relatively small industry manufacturing them had come up with a way to profit from making alternatives.

Unlike CFCs, which presented a relatively clear set of problems with definable solutions in a limited time frame, the world's growing production of greenhouse gases presents infinitely more complex challenges. The long-term scale of climate change makes it an issue that politicians can more easily duck than deal with. The scientific evidence involves massive efforts of big science institutions all over the world and is much more uncertain. Corporate profits and losses are not restricted to a single industry but involve every part of the public and private sector throughout the world. As we have seen elsewhere in this book, narrow corporate interests have continued to undermine the process of scientific review, making the development of sound policy all the more problematic. Still, despite these obstacles, some visionaries are providing us with options for avoiding what otherwise appears to be an inevitable global train wreck.

As it commonly occurs in the stratosphere—a region that extends from 20 to 50 kilometers above the earth's surface—ozone is an invisible gas consisting of three atoms of oxygen. More kinds of light are available here than at the earth's surface. Wavelengths of visible light range from about 700 nanometers, which we see as red, to 400 nanometers, which looks vaguely purple or violet. Shorter-wavelength radiation than what the human eye can make out is called ultraviolet (UV)—meaning "beyond violet." (A nanometer is one-billionth of a meter, written scientifically as 10^{-9} m.) When trillions and trillions of molecules of ozone are combined in the stratosphere, they work like a gigantic chemical sunscreen that filters the amount and types of UV radiation from the sun that hits the earth's surface. In the stratosphere, ozone absorbs most shorter, invisible wavelengths of UV radiation of 290 nanometers or less and allows some of the longer, less deadly wavelengths—UVA and UVB—to penetrate to the earth. At 290 to 320 nanometers, UVB has the ability to damage components of living cells, including our DNA. UVA, at 320 to 400 nm, is less harmful and goes right to the earth's surface.

We receive a very small amount of the sun's total ultraviolet radiation as direct visible light and as diffuse sky light that has been scattered by the atmosphere. Green growing things obtain their energy, and humans obtain vitamin D and a pleasant mood, from the right amounts of these rays. Too much UV dries up crops and injures animals and produces sunburn, eye and immune system damage, and skin cancers in humans. All living things, whether plant or animal, learn to tell time internally through exposure to light.

As with many things in life, there are good and bad places for ozone to reside. In the lower atmosphere, where people live and breathe, both naturally occurring and human-made ozone can damage living cells. In the upper atmosphere, naturally occurring ozone keeps the earth from admitting too much UV to its lower reaches. Small changes in what goes on far above the earth can have major impacts on how much and what type of UV reaches the surface. UVA radiation that reaches the earth's surface can interact with common gases of sulfur and nitrogen and break the paired bonds that hold these gases together. When an electron is taken from another molecule, it becomes reactive. Once released from these bonds, oxygen floats off in a series of solitary, floating free radicals.

What are free radicals? These are the thugs of the chemical world and can be dangerous compounds wherever they are found. Within a second of their release from CFCs, atoms of chlorine seek out ozone, which is naturally reactive to begin with, to form chlorine monoxide. In a single year, hundreds of thousand of molecules of ozone would be destroyed in this manner, creating inherently reactive radicals.

CFC molecules are combinations of chlorine, fluorine, carbon, and sometimes hydrogen atoms developed in the 1930s for use as coolants in refrigerators and as industrial solvents. In the lower atmosphere, CFCs are basically inert. Not much causes them to change composition, and for this reason CFCs were widely believed to be both safe and practical. But when they float up to the stratosphere, where they are exposed to stronger ultraviolet rays, decomposing molecules of chlorofluorocarbons release atoms of chlorine. Each chlorine atom can destroy tens of thousands of molecules of ozone. In work for which they would receive the Nobel Prize two decades later, F. Sherwood Rowland and Mario Molina in 1974 reasoned that these widely distributed compounds would directly attack the very material that kept the earth in balance—its critical layer of ozone that serves as a global sun shield.

Trained in physical chemistry, Rowland and Molina were outsiders to the specialty of atmospheric modeling. They devised a model that predicted how chlorine atoms would affect ozone, which was published in the prestigious journal *Nature* on June 28, 1974. At a press briefing some three months later, they said frankly that the 800,000 tons of CFCs produced each year amounted to a grand experiment with the earth and its inhabitants. Several hundred thousand tons of chlorine would be released into the stratosphere within thirty years, destroying up to two-fifths of the ozone layer and wiping out much of the earth's protection against chronic sunburn.

At the time, half of America's households had some form of air-conditioning. As a coolant, CFCs were much safer than what they replaced. The market potential was vast. Allied Chemical Company and DuPont anticipated worldwide revenues of several billion dollars each year from CFC products. The idea that CFCs could threaten the planet was not accorded much seriousness by the chemical manufacturers. It was pretty easy for industry to paint Rowland and Molina as two hare-brained scientists out to deprive society of progress.

A large, handsome bear of a man, nearly six feet five inches tall with a wry sense of humor, Rowland was unstoppable in his efforts to relay his findings to scientific and lay audiences. Molina was no less intense or determined.

If the earth's ozone layer were to be destroyed, life as we know it could not go on. Epidemics of skin cancer and glaucoma would spread in humans. Tiny sea creatures called krill, which feed small fish, which feed larger fish and mammals, would perish from too much solar radiation. The immune systems of many living things would be compromised. The complicated ecology of our planet was in peril.

Probably the biggest shift in how Molina and Rowland's work was viewed occurred in 1976, when a much-anticipated National Academy of Sciences (NAS) report basically confirmed their science but stopped short of urging any action. "After the NAS report came out, it wasn't just a couple of crackpots crying wolf," Rowland once quipped. "It was a couple of crackpots and the National Academy of Sciences."

Even after the publication of the NAS report and the development of a proposal for federal action, Rowland and Molina remained under siege. CFCs were a multi-billion-dollar, multinational industry. EPA targeted what it called "nonessential uses" of CFCs in spray cans, which proved to be about a quarter of all uses worldwide. The Alliance for Responsible CFC Policy, consisting of big users and producers of CFCs, came into play in 1980 to head off any proposed regulations.[4] That the person named to head EPA the following year, Ann Gorsuch, regarded the idea that CFCs endangered ozone as quackery suited them just fine.

The Chemical Manufacturers Association was becoming the chief funder of research on CFCs—a fact that, as Rowland noted with alarm in 1986, hardly made for a level playing field. The availability of ready money for research and the sheer complexity of the matter made it far more attractive for scientists to call for more studies than to recommend any action. Among other things, industry-supported skeptics noted that volcanoes and chlorine from the ocean could be important sources of ozone-depleting gases. Compared to these natural sources of ozone-destroying agents, they said, humanly made CFCs were trivial.

With its leader's insistence that "evolution is only a theory" and that trees are a major source of pollution, the Reagan administration gave new meaning to the term *crackpot science.* In February 1984, consistent with the administration's diminished concern about the environment, the NAS issued a report downsizing its previous estimates of the potential size and importance of any loss of ozone. This backpedaling was fueled more by the objections of industry than by any new analysis of the data. It was also short-lived. Just as the NAS was arbitrarily reducing the degree of ozone depletion, scientists at the South Pole had become convinced that their instruments were not working properly. Over almost the entire continent of Antarctica, the levels of ozone they recorded 20 miles above the earth's surface were lower than had ever been measured. In fact, the readings were not the problem.

Here is where the "little" science of Rowland and Molina's laboratory work came face-to-face with the "big" science of polar research—and forced it to give way. The British team stationed near the South Pole was collecting data from airplanes, stratospheric balloons, and satellite-based readings—and they could not believe what was showing up on their instruments. Convinced that the ozone-monitoring apparatus had become faulty, they had replacement equipment flown out to their isolated research station. The readings from the new instruments were the same.

One reason for their suspicions was that the monitoring satellites then flying over the South Pole had failed to find any drop. It turned out that the software processing the raw ozone data had been programmed to flag very low values of ozone as possible errors. Only after the new instruments confirmed the earlier measurements did the rest of the world accept what Rowland and Molina had predicted.

The so-called ozone hole hit the news in the fall of 1984. The previous spring, at the end of the Antarctic summer, ozone levels above the pole measured 40 percent below their winter levels. This finding created a global scientific frenzy. Once the reality of these drops in ozone became known, there could be no doubting what Rowland and Molina had done. There could be no turning back from their argument that current policies had us headed on a disastrous course.

In fact, Rowland and Molina had failed to consider that the South Pole was a pretty unusual place. Months of polar darkness are quickly replaced by months of midnight suns, with intense, ozone-destroying ultraviolet exposures for twenty-four hours a day. Even then, industry scientists challenged the findings. The head of the Alliance for Responsible CFC Policy, Richard Barnett, called the South Pole measurements theoretical. A scientist with Kaiser Aluminum, Igor Sobolev, suggested that a decade-long investigation would be required. Industry's position was clear. Given the complex uncertainties of the issue, more research was needed, before policies were adopted that could lead to economic hardships.

What all these uncertainties meant for the planet could not be determined at this time. But what they meant for industry was pretty clear. Business could proceed. DuPont at that time was expanding its capacity to make CFCs in Japan and China and had just finished a new plant in the United States.[5]

Around this time, Rowland told me of coming home late one night in 1974 after spending hours running and rerunning his model to be sure of what he was finding. He said to his wife, Joan, "The work is going very well." But then he added, "It looks like the end of the world."

By 1985, the reality of the ozone hole sent shock waves rippling through industry. Governments took what appeared to be a big step in 1987 and signed an international treaty that signaled the intentions to take what were called "appropriate measures to protect the ozone layer"—the Vienna Convention. International treaties are notorious for being broad, and this one was especially so, as it did not contain a single deadline or any real commitment to do anything about the problem, beyond setting up systems for coordinating research.

Meanwhile, several new studies confirmed that the Antarctic ozone hole had grown each year.[6] Still, for every positive result, there appeared to be a negative one—or at least that was the impression created by the alliance. The complexity of chemical reactions was easily distorted by public relations gurus into an argument for doing nothing about CFCs at all. The best of intentions became a rationale for the worst public policy.

By 1986, EPA had a new chief, Lee Thomas, who learned that the current rate of ozone depletion would create 1 million additional cases of skin cancer over the lifetimes of Americans born before 1985, and 20,000 deaths from malignant melanoma. Younger generations would have longer and higher exposures. Of Americans born after 1985, about 8.3 million would eventually develop skin cancer, and about 167,000 would die from it. If current rates of CFC use were to continue for sixty years, it would ultimately create 30 million cases of skin cancer and cause 620,000 deaths.[7]

Though not disputing these figures, the alliance argued that it would cost $4 billion to invent and introduce CFC substitutes. The EPA's policy analysts countered that the costs of *not* stopping ozone depletion ran 400 times as much—$1.3 trillion.[8]

You didn't need a degree in atmospheric chemistry to figure this one out. The costs of revamping the industry paled next to the human-health costs of continuing to use CFCs. And this was only for skin cancer. Higher levels of ultraviolet light also affected agriculture, fisheries, and the human immune system and eyes. None of these costs had yet been calculated. At a conference held with the United Nations Environment Programme in March 1986, Thomas took on those who insisted that because there were scientific uncertainties, nothing should be done:

> In the face of all this scientific uncertainty, one might ask, why has the EPA embarked on programs to assess the risks and to decide whether additional CFC regulations are necessary? Why not simply adopt a wait-and-see attitude and hold off a decision until depletion is actually confirmed? Let me address this question squarely. EPA does not accept as a precondition for decisions an empirical verification that ozone depletion is occurring. Several aspects of the situation suggest we may need to act in the near term to avoid letting today's "risk" become tomorrow's "crisis."[9]

Thomas concluded by reading from a NASA report that mirrored what Rowland and Molina had been saying for more than a decade: "Given what we know about the ozone and trace-gas-chemistry climate problems, we should recognize that we are conducting one giant experiment on a global scale by increasing the concentration of trace gases in the atmosphere without knowing the environmental consequences." Some firms directly involved with CFCs figured out that being first and best in efforts to promote their alternatives could be profitable in the short run, and convey a certain positive aura. Convinced by the growing scientific evidence, the leaders of DuPont became staunch advocates of the Montreal Protocol and promoted production practices and innovations that made it possible for global emissions to drop even more quickly than anyone working on the original treaty had dreamed possible. On the tenth anniversary of the protocol, in 1997, the Clinton administration's EPA conferred Best of the Best Awards to Coca-Cola Company, DuPont, IBM, Lufthansa, Mitsubishi Electric, Nissan, and the U.S. Department of Defense—a total of seventy individuals, associations, corporations, and military organizations from around the world.[10]

By 1988, Rowland had found a number of surprisingly astute pupils, including Senator Al Gore and Britain's prime minister, the former chemist Margaret Thatcher. After winning her third term, Thatcher, who had long disparaged environmental campaigners, made an about-face. In a speech to the Royal Society, she uncharacteristically declared the protection of the environment one of the great challenges of the century.

The mainstream had shifted and now ran directly under Rowland's and Molina's feet. Within the decade, they had received several international kudos for their efforts, including the Nobel Prize in 1995. Their work had led directly to the global treaty called the Montreal Protocol, which was approved in 1987. Unlike the Vienna Convention, this treaty bound nations to taking actions on a specific timetable. Industrialized countries would phase out the production of CFCs and fund developing countries to acquire newer materials to replace CFCs. Ironically, if their early work had been heeded in 1974 and voluntary actions had been taken then to stop releases of CFCs and create safer substitutes, there would have been no need for the protocol and several million cases of skin cancer might have been avoided. But their theory, and even some of the early findings confirming it, were no match for the political clout of the growing automotive air-conditioning business or for the confusion generated by the public relations attacks on their science.

What lay behind Thatcher's sudden change of heart that made the Montreal agreement possible? Cynics allege that she simply read the political tea leaves of rising environmental sentiment. She was also well aware that British firms had ready alternatives to CFCs and would profit directly from selling these much more costly substitutes. Britain's ICI Chemicals had a perfectly good compound available for about $4 a pound, versus $0.20 a pound for CFCs. I like to think, in addition, that Thatcher's scientific training had not entirely abandoned her.

This story of how the planet finally took action to get rid of CFCs provides an important lesson. Three things were required: the finding of the ozone hole provided sufficient proof that the planet faced a grave and imminent danger; industry had found a way to profit from making major changes in production of the source of the danger; and governments saw that the costs of persisting were much heavier than the benefits of acting. Only when all these were in place did actions to phase down CFCs really begin.[11] Within the decade, concentrations of several major CFCs in the stratosphere had begun to fall.

If coming up with measures to protect the ozone layer from CFCs proved challenging, the question of how to deal with greenhouse gases presents a bigger and harder set of issues. With CFCs, the problem was fairly straightforward. For greenhouse gases, there is no such clarity.

Although the evidence that greenhouse gases are contributing to global warming is clear—even to the administration of President George W. Bush—there is no consensus on how best to respond. Every part of modern and primitive society is involved in the generation of carbon dioxide. Every day, 6 billion persons burn fossil fuels, wood, or other organic matter that contributes to greenhouse gas formation, admittedly in very unequal amounts. Those of us in the developed world hurl nearly an SUV load of carbon into the atmosphere every year, while those from the poorest countries barely launch the equivalent of a tire. Exposures in the poorest countries, though much lower for each individual, can subject women and children to smoky fumes indoors that may be as toxic as those of cigarettes. Thus it is not a matter of one global industry that stands to gain when it makes a breakthrough, but of thousands that will incur extra costs of unpredictable size.

Some of the industries that stand to suffer the most are also the biggest and most powerful. Coal is the most widely available fossil fuel today, and China has most of the world's proven coal reserves. The average coal-fired power plant wastes more carbon than it burns efficiently. While some technologies are being promoted that would eventually inject or otherwise store carbon dioxide deep in the earth, and others are using carbon dioxide to force oil out of its wells, at this point coal remains the cheapest—and the dirtiest—way to get energy to the most people.[12] Following a tradition of naming industry-sponsored organizations for what they oppose, the Global Climate Coalition (GCC), the International Climate Change Partnership (ICCP), and the Climate Council have all formed to defend the bottom line of current fossil-fuel producers.

In an effort to understand how scientists went about measuring carbon dioxide and other by-products of fossil fuels, I once climbed to the top of the 400-foot-high monitoring tower on Mauna Loa—a 14,000-foot volcano in the middle of the Pacific Ocean. I exhaled gently into the monitor. That day the air I inhaled could well have contained contaminants smaller than 2 microns that had been tracked from specific smokestacks in Beijing thousands of miles away. The stratosphere nine miles above me undoubtedly contained molecules from the exhaust of some of those 1973 Fords that Lee Iacocca said would be hard to improve on; a trace of the pollutants Mary Amdur blew onto her Guinea pigs that Fourth of July weekend in 1950; and surely a good number of carbon dioxide molecules released by coal stoves in London in 1952 and steel mills and the zinc plant in Donora in 1948. My exhaled breath joined the millions of tons of carbon dioxide created in Donora more than fifty years ago, ending up as part of the upper stratosphere, adding insulating weight to the growing amount of warming materials now swirling around the planet. We don't know all the things that were contained in that smoke, but we do know that the molecules of carbon dioxide that those fires released will remain for nearly a century.

A lump of coal that my grandfather shoveled down the coal chute outside the family furnace in 1948 would have landed in the basement right under where my grandmother usually stayed in bed. At night my grandfather would have gone down into the cellar to fire up the furnace, and that lump would have burned completely, yielding water, carbon dioxide, sulfur, mercury, lead, a number of sticky, smelly compounds, and some heat and radiation. The heavy metals and larger particles would not have gone far. But the carbon dioxide gases are still up there.

The debate over whether global climate change is happening is over—if, that is, you are willing to accept the opinion of the 2,000 scientists who were commissioned to review the question for the United Nations and the most recent independent assessment of the National Academy of Sciences. Warming is here now; thousands of independent measurements all show a positive trend. Over the last century, average land surface temperature rose about 1.0 degree Fahrenheit, and nearly 4 degrees in some Arctic tundra. It's a big thing to change the planet's average temperature by a single degree. We know that atmospheric concentrations of carbon dioxide, the most common greenhouse gas, are 30 percent higher than before the industrial revolution. The United States releases more carbon dioxide by far than any other nation—nearly one-fifth of the total emissions of greenhouse gases.

It is not clear what all this will mean. The tundra of Siberia may become more hospitable. Fewer Siberians will freeze to death from lack of heat, as they now do. But a number of small island nations will disappear as sea level rises. Rising sea levels may create new resort communities and devastate low-lying coasts. Not everybody can afford to build tall brick barriers or dikes to keep the sea out. Water supplies and hazardous waste sites will have to be moved away from salt-water intrusions.

Some astonishing effects are already apparent. At the end of March 2002, the northern section of the Larsen B ice shelf, a floating ice mass larger than the state of Rhode Island, shattered and separated from eastern Antarctica, leaving a massive drifting plume of thousands of icebergs in the Weddell Sea. The ice sheet had melted faster in a single decade than in the entire century before. The last group to try to walk to the North Pole ended up having to sail over it because so much melting had occurred. In the Himalayas today hundreds of thousands of people are at risk of sudden flooding from melted glaciers and may lose their water supplies and their homes. Moosely Seconds mountaineering store in Grand Teton National Park has had two consecutive "global warming sales" with rock bottom prices on ice axes and crampons because the usual summer snows had disappeared early in the season.

Tom Mangelsen, the award-winning wildlife photographer, returned in July 2002 from a two-week naturalist expedition to the northern polar latitudes with an unusual number of photos of swimming polar bears and cubs. At the top of the Arctic food chain, polar bears target seals from their strategic vantage perched on ice floes during the brief but intense summers. For the past few years more and more bears are swimming, as they have been running out of ice from which to hunt their prey. More are starving. Fewer are storing enough fat to survive.

Further south, grizzly bears are ranging to lower altitudes looking for food in Glacier National Park because the berries they normally eat in early summer are drying up under the hotter sun. Tree and plant species are growing in more northern latitudes than ever before. The world's ecosystems are under stress, whether forests or coral reefs. Of course, we can't be certain that these events are due to a warmer planet, but nobody has come up with a convincing alternative explanation.[13]

What can we do to reduce these impacts? That is the multi-billion-dollar question on which the future rides. Talk about a big science problem. There probably is none bigger or more complicated than what to do about climate change and its impacts.

One of the most amazing feats of modern engineering took place in 1999, when the National Park Service moved the Cape Hatteras Lighthouse nearly half a mile. The nation's tallest brick lighthouse was about to fall into the sea. Located about a quarter mile from the shore when it was built in 1870, by 1987 it sat only 120 feet from the shore. After months of planning, steel support towers with hydraulic jacks were placed at the base of the lighthouse. The 200-foot-tall building was then lifted 6 feet and given steel support beams as a temporary foundation. One hundred hydraulic jacks on rollers slid slowly along track beams to move the massive structure along a specially constructed level roadway. From June 17 until July 9, 1999, the lighthouse inched along, arriving three weeks ahead of schedule. Again, the beacon is now 1,600 feet from the shoreline, and open to the public.[14] If the twentieth-century's rate of erosion holds, it should be safe for another hundred years.

This incredible engineering feat shows us what we can do if resources are not constrained. It is too bad we cannot simply redesign how we use carbon as deftly as those engineers managed to move the lighthouse.

Adding smoke, if not fuel, to the fires on global warming is the highly polarized atmosphere in which the serious scientific issues are being addressed. Because every nuance receives such magnified attention, there is a tendency toward exaggerated arguments on all sides.[15]

In the summer of 1997, the editors of *Lancet*, the world's oldest medical journal and also one of the world's most innovative, instituted a rapid review policy for important new work. Under this policy, I proposed a global assessment of the public health impacts of continuing to burn fossil fuels as usual. Given that the International Framework Convention on Climate Change was scheduled to take place in Kyoto in December 1997, I thought it important to get this information into the hands of world policy makers in a form that would get their attention. *Lancet* typically rejects flat-out nearly 90 percent of what gets sent to it. Still, the prospect was too tempting and the subject too important not to try.

Here's what a team of WHO, EPA, Harvard, and WRI experts did to answer the question. First, we divided the world up into nine areas. For each of these regions, we relied on estimates created by the United Nations of what constituted normal practices—business as usual. In each zone, we looked at current technologies for energy, transport, industry, and housing. Next, we estimated the carbon dioxide emissions and principal pollutants generated by each of these sectors every year and modeled how they would release pollutants and greenhouse gases. We also estimated how much these releases could be cut by adopting reductions being proposed by various groups, such as the European Union, over the next two decades. Finally, we put all this together to estimate how many lives could be saved by 2020.

Within a month, we had our answer, in 6,000 words and ten tables. If the world continued to burn fossil fuels at the current pace, by the end of the second decade of the twenty-first century, 8 million avoidable deaths would occur solely from controllable exposures to particulate air pollution. In response to our analysis, *Lancet* sent us seven pages of single-spaced criticisms, to which we were given a week to reply. We needed it. Our paper appeared in November 1997, and made it around the world.[16] Normally, general sessions of international meetings are not that well attended. In much of life, timing can be everything. Speaking just before Vice President Al Gore at the Kyoto Conference on Climate was the president of Nauru, which is a small island nation threatened immediately by rising sea level and storm surges. Nauru's president referred to our *Lancet* study on the hidden health benefits of climate policy in the general plenary session as a wake-up call to the world. The next day, the work had front-page coverage in China, Japan, Mexico, France, and Germany, including positive mention in British media by the deputy prime minister of England, John Prescott, and in Brazilian papers by Fabio Feldman, a senator who directed one of Brazil's largest environment agencies.

A few weeks after this whirlwind, while in Xinji, China, I got an urgent fax from the *Lancet*. They had received what appeared to be a very serious set of charges about our paper, claiming that our data were flawed and our analysis was wrong. I did not recognize the name of the person making these charges. He had never published a single peer-reviewed article in the field of environmental health. His recent work consisted of letters to the editor and other attacks on groups that had estimated the public health impacts of air pollution, including the Toronto City Health Department and the Canadian national government. The papers this unknown critic relied on came from something called Citizens for a Sound Economy Foundation. He argued that there was a level below which no health effects could be expected—a threshold for air pollution. I spent more time rebutting what he wrote and persuading the *Lancet* that this was not a serious scientific critique than my team had spent in responding to the original peer reviewers' substantive comments. For anyone who wants to read about it, this entire exchange, which the *Lancet* declined to publish, can be found on the Junk Science Web page, whose creators thought it suitable for posting.[17]

Based on what we had presented at the Kyoto Conference, I was appointed an official member and lead author of the UN's Intergovernmental Panel on Climate Change (IPCC)—one of more than a thousand experts to work on what became the Third Assessment Report on Climate Change. I came into this as a real outsider—a public health expert among climate modelers and economists. The Energy Modeling Forum had been noodling around the issue of estimating the costs of changing policies for years. I asked Jae Edmonds, an acknowledged sage in the field, whether anyone had ever estimated the public health costs of continuing things as they were and the benefits of reducing this damage through climate policies. His first reply was, "In our models, we assume that all public health benefits have already taken place. They are kind of like hundred dollar bills lying on the ground. They do not exist, because if they did, they would all be picked up."

This was an interesting way of looking at the issue. In other words, if there were any negative impacts on public health from current fuel and energy policies, these would be getting fixed by other means. This was a polite variant of "Go away, kid. Don't bother me."

In fact, the final report of our IPCC group did allow that the economic benefits—in terms of avoided health damages—from reducing greenhouse gases could be as large or larger than the costs of implementing those policies, especially in developing countries. It even quoted the estimated 8 million avoidable deaths that would occur if nothing were done to change current policies. That may not seem like much, but that acknowledgment represented a huge victory.[18]

There is a growing recognition that the increased level of monster storms, tidal surges, downpours, droughts, searing temperature, and windstorms we are seeing today is only a dress rehearsal for the global warming to come. My argument, within this much bigger issue, is that there are also public health benefits now to reducing carbon emissions. Among the talented people who have joined with me to make this case is Luis Cifuentes, an associate professor at the Catholic University of Chile, with whom I have been working for more than five years. His graduate and postdoctoral adviser at Carnegie Mellon University was Lester Lave—the same fellow who laid down the basic methods for conducting air pollution epidemiology in 1970. Cifuentes has come up with a way to calculate the health benefits from adopting specific energy and transport technologies in his country's capital city of Santiago.[19] In the past few years, particles in the air of Santiago have been lowered through cleaner engines in buses and trucks, lighting that yields more brightness for less energy, recycling waste heat, and the use of cleaner-burning fossil fuels along with some renewable forms of wind and solar energy. Every single one of these technologies is available today. We need not wait for Buck Rogers to fly in off his spaceship to put them to work.

Nelson Gouveia of the University of São Paulo Medical School has teamed up with Tony Fletcher of the London School of Hygiene and Tropical Medicine to quantify the effects of air pollution in São Paulo on the elderly and the young. São Paulo is one of the most congested cities in the world; the region's 8 million cars use just five main roads to move 16 million people around. Gridlock is too tame a term to describe the result. Rats raised in dirtier sections of the city have lungs that look like those of smokers. The gases that are contributing to global warming here are the same ones that are killing people on the ground.

In Manhattan, George Thurston of New York University has led efforts to show what current levels of pollution mean for this resilient city. More have perished from air pollution in the 1990s than died in the tragedies of the World Trade Center. Cifuentes, Gouveia, Thurston, and I, with Victor Borja-Aburto from Mexico, pooled our resources to produce an estimate that was published in *Science* magazine in August 2001. São Paulo, Santiago, Mexico City, and New York City together have a population of 45 million. Adapting the framework from the *Lancet* study, we estimated the number of lives that could be saved under some fairly conservative assumptions.

We did not take into account such efforts as reducing the indoor use of dirty fuels—which expose millions of women and children to stifling levels of pollution—or assume that any new technologies would be widely adopted. Simply through increasing the energy efficiency of fuels, using cleaner fuels, or using more solar and wind power, some 64,000 premature deaths, 65,000 cases of chronic bronchitis, and 37 million days of restricted activity or work loss could be avoided *in these four cities alone* through 2020. Of the twenty-four cities in the world today with populations approaching 10 million, eighteen are in rapidly developing countries. The numbers of lives that could be saved and helped through cleaner energy run in the millions. The costs of making these changes will be very large as well, no matter how they end up being calculated. But too few people stop to consider the cost of *not* making them.

Are we sure of these numbers? Of course not. We make such estimates precisely because we do not want to have to wait to be proven correct. Before we attempted to come up with some assessment of the public health impacts of fossil fuels, no one had done it. Could we be wrong? Sure. But we are not wrong to ask these questions. We create these assessments for only one purpose—to provide some way to gauge the potential scale and scope of what is involved. Of course, the impact will vary with whatever technologies are adopted. In the United States, the Union of Concerned Scientists has calculated that if new cars achieved an average of 40 miles per gallon, rather than the 27.5 they get today, this would be equivalent to removing 44 million cars from the road.[20] Adopting currently available and more efficient fuels in our homes and factories could lower total production of carbon dioxide by 60 percent and keep 53 million tons of heat-trapping gases from reaching the upper atmosphere.[21]

Each year about 6 billion metric tons of carbon are released into the atmosphere of this planet. In the modern industrial era, concentrations of warming gases—carbon dioxide, methane, and nitrous oxide—have risen more than in the past ten thousand years. In 1997 the Kyoto Conference on Climate Change provided a rhetorical victory that included a broad commitment from developed countries to reduce their emissions by 2012 an average of 5 percent below what they had been in 1990. Because the United States experienced tremendous economic growth in the past decade, in our country this amounts to its lowering emissions more than 25 percent below what would have occurred without the treaty. For developing countries, the treaty required no specific actions but provided ways for them to get funds and technologies to promote efficiency and cleaner fuels.

Like many global treaties, including the first one created on CFCs, the Kyoto Protocol offered no way to check compliance and no way to compel actions of major carbon users. In fact, the Europeans face a less difficult task in reducing carbon than does the United States, which had a booming economy until recently. Germany can claim credit for reduced total emissions that have come about solely from the reuniting of East and West, and England gets credit for shifting from coal to its vast stores of North Sea natural gas. These are one-time events that cannot be duplicated anywhere else.

In an adroit political move, President Clinton in 1997 signed the Kyoto treaty but never submitted it for ratification, knowing full well it was dead on arrival. The same year the treaty was signed, the U.S. Senate had voted 97 to 0 against the ideas it contained. President Bush has announced that he has no intention of giving any support to the Kyoto Protocol but has not come up with a concrete alternative.

About the new Bush policy, the British foreign minister exclaimed, "[The Kyoto Protocol] was signed up to by every single nation on earth, and if America now tries to walk away . . . I think this is not just an environmental issue, it's an issue of transatlantic global foreign policy." In fact, this statement is not correct. Developing countries could sign the treaty but were not obligated to take specific actions.

The chorus of criticism of the U.S. current policy is pretty broad and includes some of the worst national actors in the world. Australian Environment Minister Robert Hill said the collapse of the Kyoto Protocol would be "a major step backwards." This is an ironic position, given that each person in Australia already emits more than five times more carbon dioxide than each person in Asia.

A serious gap in leadership exists on this issue today in many countries, with lots of cheap talk and accusations being traded and little being done. A number of other countries and 116 U.S. cities, counties, and states have taken steps to reduce their production of carbon. In Canada, where each person sends the equivalent of a light-duty truck load of carbon into the atmosphere each year, the federal government has agreed to reduce greenhouse gas emissions by more than 23.7 megatons—more than 10 percent of current levels—by the end of this decade, and has committed $1.1 billion to the effort. Unfortunately, the fellow who announced this policy, Canadian Prime Minister Jean Chretien, is likely to be out of office long before the promise comes due. These Canadian assurances may turn out to be as empty as those made by the United States at Kyoto.

All these statements move in the right direction, but few of them represent any serious actions, and none, even at the national level, can achieve much real change by itself. Life would be simple if all we needed to do was figure out how to get the world's biggest users of carbon dioxide to shape up, or move the 100 million people who live on the seacoasts a half mile inland, like the Cape Hatteras Lighthouse. But the problem is much more complex than that. It requires basic reengineering of much of our lives. The good news is that some efforts to reduce greenhouse gas emissions are well under way, in places as far-ranging as Oberlin College, the "green" ranch of President George W. Bush in the west Texas desert, one of New Jersey's largest and most profitable hospitals in Hackensack, and several national parks in the United States.

Oberlin College features a new environmental studies building that incorporates a number of innovative green technologies, sometimes exporting power to the local utility. With 3,700 square feet of photovoltaic panels on its roof, the building uses 80 percent less energy than standard academic buildings and also employs a natural wastewater treatment facility of plants and ponds, with a wetland for natural storm water management.

George and Laura Bush's Texas ranch recycles wastewater and collects rainwater. Buildings on the ranch employ a passive solar design that naturally absorbs winter sunlight, warming the interior walkways and walls. Underground water at a temperature of 55 degrees is piped through a heat-exchange system that warms the interior in winter and cools it in summer.

One of New Jersey's largest hospitals, Hackensack Medical University, spearheaded by two of its biggest donors, Don and Deirdre Imus, has embarked on a major program to revamp its use of cleaning products. The hospital has switched from those that require lots of energy to produce, such as toxic organochlorine compounds, to those that are less toxic and more easily generated.

In several national parks, Sophia Wakefield has worked her magic hands. Yellowstone, Grand Teton, Yosemite, and Mount Rainier, among others, use green cleaning products that naturally degrade, are less harmful to workers, and do not impact the environment as harshly as do the materials they are replacing. It takes less energy to produce these materials, and they weigh less heavily on the environment when used.

Several other, much bigger and brighter lights on climate actions can be found in both the public and the private sectors today. Tired of the meaningless gestures from the last few federal governments, some in industry, such as British Petroleum (BP), are organizing themselves to reduce carbon and convert to healthier forms of energy and transportation. In March 2002, BP announced that it had reached its self-imposed greenhouse gas emissions reduction targets of 10 percent below 1990 levels eight years ahead of schedule at no increase in costs at all, reducing total emissions more than 9 million metric tons since 1997. Although these claims are welcome, we still need some sort of independent way of certifying that these advances have taken place. Because there is currently no such system on the horizon, the field is ripe for those who might wish to overstate their efforts.

The International Organization for Standardization (ISO) sets standards for everything, from indoor air supply to the quality of bricks used in construction, relying on thousands of expert committees. In 1996, ISO 14000 created a suite of environmental practices for "socially responsible investments," including commitments to energy efficiency and recycling. By 2000 more than 22,000 had signed on. Paul Faeth, who directs WRI's economics program, pointed out that of the world's hundred largest economic actors, fifty are corporations and fifty are countries.

Corporate policy can be even more important than government actions in determining what happens to the planet on which we live. For example, in 2002, eight major carpet manufacturers agreed with EPA to eliminate landfill disposal and incineration of used carpet by coming up with a recycling program that works. This could keep 2.5 million tons of waste out of garbage heaps and put it back into circulation in other forms.

Possibly the most inspiring words and deeds on new sources of lighting, heating, and energy are those that are returning to the original sources, so to speak. A growing number of churches and synagogues are relying on energy that comes from, and returns to, the earth itself. These sources of energy include renewable ones like wind and solar, which emit no carbon at all. I serve on the board of the Coalition of Organizations on the Environment and Jewish Life, as part of the National Religious Partnership on the Environment, which promotes greener energy for its members and within various institutions. They have even created entertaining teaching tools for the candle-ridden holiday of Hanukah, called "Let there be renewable light!"[22]

Before the Kyoto Conference in 1997, the General Convention of the Episcopal Church, USA passed a resolution calling for energy efficiency in houses of the Lord. The Episcopal Power and Light ministry has spread like a spiritual wildfire, having begun with the Regeneration Project, a San Francisco–based public charity, a project of the Tides Foundation. Many of the participating churches have chosen to purchase renewable energy from Green Mountain Energy Company. Founded in 1997, Green Mountain is now the largest and fastest growing U.S. provider of less-polluting electricity generated from wind, solar, water, geothermal, biomass, and natural gas. Right off the Pennsylvania Turnpike sits one of the newest wind farms in the United States, in the town of Garrett. Two of the largest solar facilities are in northern California.

Much more than rhetoric or theological punning is involved in the religious organizations' commitment to renewable energy. By combining the purchasing power of churches and their congregations, Episcopal Power and Light receives a rebate of $35 per household from Green Mountain. In Clayton, California, Saint John's Episcopal Church has earned upwards of $4,000 from having their members sign up to switch their household energy supplier.[23]

Sam Wyly, the populist leader of Green Mountain, explained that he got into the business of renewable energy for a very simple reason. He wanted to see the stars again. In Dallas, Texas, where he lives today, he admitted, "The stars at night are no longer big and bright. A brown haze, which I used to see on trips to Los Angeles or Taipei, has slowly been creeping across the Dallas sky. The problem has hit home."

Wyly is a shrewd capitalist who is the intellectual offspring of what he describes as the peculiar marriage of Adam Smith—the father of economics—and Rachel Carson, the mother of environmentalists. He had this to say about the imbalance of resource usage by Americans:

> The generation of electricity is the single largest source of industrial pollution in the U.S. today. It is an industry dominated by dirty coal-burning plants, which have contributed hugely to climate change, acid rain, smog and an increase in respiratory illness. Air pollution represents one of the great threats to life on this planet. We Americans are less than 5% of the world's people and we emit nearly 25% of the earth's air pollutants. 2 billion of our planet's 6 billion people still don't have electricity. Will ours and theirs be clean or dirty tomorrow?[24]

He chastised Texas for spending a billion dollars a year importing coal from Montana for electricity, while failing to use sunshine, wind, and tides. Wyly's company is making money and boasts of taking the equivalent of a million cars off the road by its use of cleaner energy.

The costs of reducing pollution, whether undertaken to prevent local air pollution or to lower carbon dioxide, have to be gauged against the major benefits to public health, which cannot always be calculated financially. The interrelated efforts of energy conservation and greenhouse gas reduction, such as those at President Bush's ranch, are not, despite what Vice President Dick Cheney says, a matter of private virtue. If we depend solely on those who can afford to do the right thing and are prepared to pay extra for doing so, we will never change the course of the supertanker in which we are all riding. Today, there are 150 million American families who are the real decision makers on climate. All families anywhere make choices based on keeping their out-of-pocket costs low and their usable cash high. Unless there is some incentive to change how moneys are spent, we will never succeed in producing the sorts of change that will have any major impact on climate.

The effect on the environment has to be factored into the economic decisions we make every day. We need to find some way to make the price of turning on our lights and getting to and from our jobs reflect the full and real health and economic costs of those choices. Coal and diesel fuels are cheaper to use, because the prices we pay for them do not include the costs of the additional health-care expenses, workdays lost, anxiety spent worrying over our children's asthma or other health problems, and impacts on climate. Every ton of coal burned in Ohio means more children in Canada and the Northeast will be stricken with asthma or other lung problems. Every time Mom turns the key in the family's fat, new gas-guzzling SUV, another child downwind faces an increased threat of wheezing or being wheeled into the emergency room. If we continue to subsidize gasoline, build roads, and encourage people to drive and park their cars in our cities for less than what it costs in services, we are perpetuating a culture that is dependent on pollution.

The economists have a solution for all this, and it should be accorded much broader public discussion. We need to pay the full costs of the "bads" that result from our current use of energy and find a way to give money back to those who lack the resources to afford these costs. This means that things that contain lots of carbon, such as coal and gasoline, should cost more than things that contain less, like natural gas, solar, or wind energy, or fuel cells when they are better developed.

These changes will require what free-market economists and their conservative followers regard as the two great liberal evils: subsidies, to encourage the use of cleaner energy sources, and taxes and other penalties, to discourage the use of dirtier ones. I would only point out that the free market is not an end in itself. We should support it to the extent that it benefits the public welfare, and apply the same standard to government intervention in economic affairs. Despite the dire warnings of industrialists, the actual costs of such interventions to control local air pollution have proved much smaller than the benefits to public health from taking such actions—about ten dollars saved for every one spent.[25]

Some commentators—most recently, Bjorn Lomborg in his book *The Skeptical Environmentalist*—have argued that the money it would cost to reduce our use of fossil fuels and switch to greener energy would be better spent on wiping out world hunger or providing clean water.[26] Maybe so, but there are some problems that cannot be fixed later on, no matter how much more money we may have accumulated in the meantime. As J. F. Rischard points out in his new book *High Noon,* the problems of hunger, climate, and freshwater supply all have two things in common: They're getting worse, and they'll be considerably more costly to fix later, if at all.[27]

Among all the world's problems, climate change has a certain distinction. If we fail to deal with the prospects of what the British environmental scholar Anthony J. McMichael termed planetary overload, the fact that we had more well-fed, sated people before reaching this point will be of no consequence.[28] We are contending with a grand experiment with an *N* of one. The risks of irrevocable damage to the earth's climate are so great that waiting for certain proof constitutes a doomsday experiment.[29]

Speaking of the potential irreversibility of atmospheric change, Lester Lave, now working at Carnegie Mellon University on matters of energy efficiency and climate, has quipped, "It will only take us a thousand years to recover." Lave often works closely with those in industry, and he certainly understands that their concerns with costs are not unreasonable.

> If there were a way to lower the cost of abating greenhouse gas emissions, most prudent people would decide that they are unwilling to risk the Earth to this Frankenstein experiment. There are lessons from air pollution that give us hope. When the government told people precisely what to do, abating air pollution was expensive. It became much cheaper when government set the goals, in terms of reduced total emissions, and then allowed industrial facilities to find the cheapest way of meeting these goals.

Lave argues that global warming is a much more difficult problem, both because of the long time frame and because we need to get all of us, not just the industrial facilities, to act. Just as businesses need to have a goal and the freedom to meet it, so do consumers. Economists say that this could be done by giving each individual certificates to put a certain amount of greenhouse gases in the air. Individuals could trade these certificates, with some people selling and other people buying them. People who walk or bicycle to work could sell their certificates to those who live in the suburbs. Another alternative would be to charge users for the amount of greenhouse gases they emit. For example, a tax could be levied on every gallon of gasoline to account for the carbon dioxide that comes from the car's tailpipe. People buying electricity made from coal would pay the tax while those buying electricity generated from windmills would not.

These days, however, no one is ready to propose new taxes on anything. Gus Speth, one of the founders of the Natural Resources Defense Council, and now dean of the School of Forestry and Environmental Studies at Yale, admits that the failure to address economic issues has haunted U.S. policies on the environment for years:

> Thirty years ago the economists at Resources for the Future (RFF) were pushing the idea of pollution taxes. We lawyers at NRDC thought they were nuts and feared that they would derail . . . the Clean Air Act, so we opposed them. Looking back, I'd have to say this was the single biggest failure in environmental management—not getting the prices right.[30]

A team working at RFF, which is now the leading think tank on the issues of natural resource economics, has come up with an appealing scheme that gets well beyond taxes.[31] Here's how one really creative economist describes it:

> Fundamentally, we need to motivate the whole society—not just the well intentioned or those with the extra bucks to spend on being green. There is only way to do that. We must come up with a way to make it pay for people to use cleaner, more efficient forms of energy. The only way to do that is to create genuine, lasting incentives that make it cost more to release carbon into the air. We have to put more money back into people's pockets if they spend for cleaner fuels.[32]

People respond to prices—if it costs more to burn dirty coal in a boiler in your apartment building in Taiyuan, China, then people will switch to cleaner forms of energy. The good news is that both corporations and households respond to *future* prices as well as current ones. This means that promising to make it cheaper in the future can get people to change what they are doing now.

When I asked how you could possibly change the prices of things without putting a tax on them, he stressed that taxes would never fly.

> Fuggedaboutit. The key is to return the revenues—or at least most of the revenues—directly to households in a highly visible way. We need to give people a check, or a coupon. That is what excites people. This is not rocket science—it just takes a little imagination and some leadership to create enthusiasm for such an approach. Besides, someday soon, the government is going to run short of revenues because of all these tax cuts. The next president is going to have to deal with it. If we come up with a large revenue stream in the service of the planet—most of which is returned in a populist fashion—this could be a natural solution to what would otherwise prove our undoing.
>
> The bottom line is that we either come up with a way to pay today for the full costs of what we are doing to the planet, or we will all pay tomorrow in ways that we cannot even imagine.

10

DEFIANT FIGURES

The right to search for truth implies also a duty.
—ALBERT EINSTEIN

WITHIN A MONTH after fire destroyed Hitler's headquarters at the Reichstag on February 27, 1933, the German government began sending truckloads of communists, Social Democrats, students, and others suspected of being behind this attack into "protective custody" on the grounds of the former munitions factory at Dachau in southern Germany. This was Germany's first concentration camp. From its opening on March 22 of that year till its closing on April 29, 1945, the camp registered the arrival of 206,206 inmates. The number of deaths officially recorded at Dachau is 31,591. This total does not include the thousands who perished in massive evacuations and death marches before they ever arrived, those singled out by the Gestapo for what was euphemistically termed special handling *(Sonderbehandlung),* or those who were killed because they were Soviet prisoners of war.[1] The true toll may well be four times as much. For all their meticulous record-keeping, the Nazis never did manage to tally all the victims of their horrific industrial-scale murders.

Today, public transportation makes the trip to Dachau quite simple. From Munich's bustling, cavernous main train station, you take a twenty-minute ride on the S-bahn, walk directly out of the local subway station to the Gedenkstatte parking area and hop on Bus No. 724, which takes you right to the camp's notorious gates.

During the brief inspections by the International Red Cross when the camp was operating, baskets of flowers would stream from wooden posts that normally served as gallows. Official visitors would stroll past the special area reserved for priests, political prisoners, Jewish doctors, and high-ranking dissidents. In the open, central parade grounds, inmates were forced to remain motionless every day, sometimes through the night, until they had all been counted by the numbers tattooed on their arms. The incinerators sat well out of view.

In the vast remains of the camp, across a huge empty space the size of several football fields, a small bronze man perhaps five feet high stands alone in a corner just outside the hidden ovens. When Dachau was in business, the thousands of men imprisoned there were required to have their heads bowed and covered at all times. Prisoners were not allowed to look guards in the eye. They could be beaten or shot if they failed to stand at attention, feet together, hands at their sides and visible at all times, coats fully closed.

The head of this small statue is upright, staring straight ahead with an unyielding expression. Both hands are set deep into the pockets of his overcoat, which is not buttoned at the top. The feet stand apart with toes pointing outward. This small monument is called simply The Defiant Inmate. On the base of the figure is written in German:

> By honoring the dead, we warn the living.[2]

Few of us have ever lived in a world where simply having our hands in our pockets or wearing our coats unbuttoned was a heroically subversive statement. I certainly never have. But the unique and incomparable evil of the Holocaust does not remove it entirely from the human realm. The Defiant Inmate is, after all, doing nothing extraordinary. He is only trying to live with a minimum of dignity. It's his environment that has turned this mundane behavior into a life-threatening action.

The world is very generous in providing situations in which doing what seems natural, decent, and inoffensive; pursuing one's work; or seeking to discover something true and useful can require a surprising degree of courage and tenacity. This book has shown how, time and again, the ability to broadcast warnings about environmental health issues has been repeatedly hampered. Scientists brought in to count and measure ties between the environment and health often faced daunting restrictions. Each time I looked beneath the surface, I found worrisome surprises.

For years, few people tried to address the connections between health problems and poor environmental or workplace conditions. Mary Amdur was one of the first to take the risk. A Pittsburgh girl who made good in the Ivy Leagues at Harvard, in the 1950s Amdur resisted intense pressure to suppress her pathbreaking work measuring the hazards of slow, regular exposures to acid aerosols. Forty years later, she still recalled "feeling as though hands were about to close around my neck" when two strange men confronted her in an elevator and signaled that she should not publicize the results of her work.

In the early years of the EPA, riding a broad tide of public support, agency officials held lobbyists for the car firms at bay and pressed ahead with efforts to reduce pollutants from cars. In the 1980s and 1990s, a kind of arrogant bravado fueled efforts to dismiss scientists who warned of the hazards from some ordinary forms of environmental pollution. After all, our society was growing economically. Many people were living longer and better. Worrying about small, not well characterized problems that might arise from the way our improving lives were organized seemed a pretty remote and indulgent set of concerns.

Throughout these mostly booming decades Herbert Needleman, who quantified the subtle and insidious effects of lead on children's nervous systems, survived highly personal attacks intended to discredit his work and paint him as a renegade and biased researcher. He stood up to the dominant view that children's problems could mostly be blamed on bad mothering, poor housekeeping, and lousy parental role models. Sherwood Rowland and Mario Molina persevered, too, despite industries' campaigns to label them as antigrowth extremists for their work showing how CFCs could destroy our planet's ozone layer. They were vindicated for all time by the Nobel Prize Committee. It helped a great deal that the evidence showing they were right could not be argued away or made to disappear. The efforts of Victor Borja-Aburto, Carlos Santos-Burgoa, Lester Lave, and Mary Amdur to protect people's health should not have been controversial—should not have exposed them to the attacks of less courageous people who were paid to discredit them.

Human health is the result of lots of things that happen to us, even before we are conceived. We know that what dads do in the months before they become fathers can affect the health and even the sex of their children. We have heard much good advice about all the things we are supposed to do for our health. Yes, we certainly should not smoke, drink or eat heavily, or indulge in bad habits and dangerous sexual practices. Obesity is a growing problem. So are binge drinking and smoking in the young today, which will surely lead to epidemics of liver disease and many forms of cancer in the future. But there are some problems that cannot be addressed by individuals, no matter how well they try to take care of themselves or how many resources they may have to do so.

Far too often, people get no chance to discover and speak the truth, because records are sealed or were never developed in the first place. In Anniston, Alabama, for years the failure to report hazardous discharges to the local rivers from Monsanto's PCB-polluting plant—later bought by Solutia Corporation—subjected thousands to potentially grave risks, the full nature of which will never be resolved. Recent proposed settlements in that case will not necessarily yield useful health information but will limit corporate liability.[3] Whatever the outcome in this case, numerous other poor, often African American, towns have been effectively wiped off the map altogether as a result of environmental contamination. In Iberville Parish, Louisiana, pollution was so heavy that Dow Chemical settled lawsuits agreeing to move out all of the residents of the former towns of Reveilletown and Morrisonville. No wonder more studies are needed. In these instances, people are not just lost to follow-up studies that are never conducted. They are also lost to their homes, schools, churches, and all other connections to their neighborhoods.

In 2001, Elihu Richter and colleagues compiled a list of fourteen instances in the United States and other countries in which public health professionals were prevented from amassing data, where data were distorted by public relations concerns, where researchers were attacked or removed from their positions after warning of hazards, or where they were blackballed or gray-listed from participating in research on the environment.[4] Toxic substances and microbes need not show passports as they circulate throughout the world today, but those seeking to study them can be denied access to the tools needed to study their impacts through a variety of sometimes insidious processes.

When all is said and done, the surprise is not how little we know about public health today, but how much we have been able to learn despite the barriers to science, the outright lying and deception and intimidation, and the political and economic pressures that continue to make it difficult to conduct studies at all. Whether we bother to look for connections between the environment and our health is not merely a matter for science. Society makes choices every day regarding what it values, and what we are prepared to try to change and spend our resources on. In the United States today, we expend more than $400 billion each year[5] treating individuals like my mother and her sister who are coping with chronic illnesses, like cancer and heart disease. Yet, we do too little to identify and control exposures that could lessen our reliance on such interventions. Recently, at a cost of millions of dollars our airports and other public places have sprouted impressive-looking, wall-mounted, cardiac defibrillators to issue life-saving electric shock to those individuals who might just happen to be in the right location when their hearts suddenly stop beating or become so erratic that they can be shocked into a normal rhythm, with a trained health professional nearby. This equipment will save lives of chronically ill individuals who frequent such public places, one at a time. No matter how efficient our health care system may eventually become, we still need to reduce the demand for such heroic measures. We can do this by making better use of what we already know about how the environment shapes the lives of all of us.

The budget for research on environmental health remains paltry compared to the sums invested in high-tech treatments and interventions. Still, money alone does not ensure that the right work gets done in the best manner. Although more federal money is being spent now on breast cancer research than at any other time in history, the proportion spent on environmental issues remains less than 1 percent.

We need institutions that can answer important scientific questions about the environment without fear of recrimination or the withdrawal of support. Few, if any, of our present institutions can provide the sort of genuinely unbiased, hard-boiled, independent science we need today. We have seen how, in an attempt to magnify the controversies and distort the findings on areas of legitimate controversy on environmental health, the public relations industry has easily exploited the natural tendencies of scientists to argue over fine details. Sometimes the fights have been nasty and underhanded. The lead industries, for example, brought suit against a Harvard pediatrician, Randolph Byers, in the 1940s, seeking to suppress his findings that children who ate lead paint chips could become poisoned. In the 1980s, the industry fomented charges of misconduct against Needleman.

Public health researchers have not always realized how easy it was for others to exploit their legitimate admissions of scientific uncertainty. We always need to know more in science. But if we always insist that we should do nothing until the damage is absolutely certain, then the only certainty is that we will cut short millions of lives and bring misery to millions of others.

For years, life, death, climate, and sex were all believed to be beyond our control. We now understand that each of these complex and exquisitely important aspects of life is subject to more human influences than have been imagined. Global warming is not just some vague and fuzzy thing that may someday make it possible to grow flowers outdoors in New York year round and improve the value of shorefront properties in Canada. It turns out that the things that are changing our climate are the same things that caused so much pain in Donora, in London, and throughout the world. Those confused swimming and starving hermaphroditic polar bears and whales in the Canadian Arctic with high levels of PCBs in their fat show that other kinds of pollution now have global reach as well. Even though diseases affect people as individuals, there are patterns of exposures and conditions that influence how these diseases arise and that can be addressed through social action to reduce or control such conditions.

The good news is that here and there, people are beginning to act on this evidence of environmental influences. People care passionately about their children and now have greater access to practical advice on how to better protect their environment, thanks to a broad range of groups who are organized about this issue.[6] More than a decade ago, the Swedish oncologist Karl-Henrik Robert[7] devised the Natural Step program aimed at using fewer toxic agents in daily life. As people have become convinced that cancer and other diseases were connected to environmental conditions, his work is no longer considered radical or revolutionary. Mitchell Gaynor, the charismatic director of the Sanford Weill Center for Alternative and Integrative Medicine at Cornell Medical School, routinely tells cancer patients to reduce their uses of suspected hazardous agents in order to improve their prospects of living longer and better.[8] Gaynor is also leading a campaign to stop the construction of what would become North America's largest cement industrial city of more than twenty buildings atop an 1,800-acre ridge in the Hudson Valley, adjacent to the heritage river site near his home. The plant would release 1.47 million pounds of particles each year from a stack that will be one of the tallest structures from New York to Montreal and would dwarf the Statue of Liberty. Deborah Axelrod, the breast cancer surgeon who wrote *Bosom Buddies* with Rosie O'Donnell, knows how to use humor to diffuse the trauma of cancer.[9] She is equally adamant in her clinical practice about the need to look beyond pills and traditional medical interventions to reduce the chance that the disease will recur. Andrew Weill counsels those who want to stay healthy to read labels of their foods and look at what is under their kitchen sinks.[10]

Efforts to recycle, renew, and conserve have moved out of the backwoods and into the mainstream. The Ford Motor Company recently announced that its vast River Rouge plant in Dearborn, Michigan—one of the great landmarks of the American industrial revolution—will now be more environmentally friendly, with energy-efficient buildings, an on-site wetlands to help process wastes, and many other innovations. As recently as 1999, a massive explosion in the outdated coal-fired power plant of the facility's 1,100-acre site killed six workers and injured fourteen. The smokestacks will be torn down, giving way to what is being billed as an ecologically designed facility. If Ford can do it, other businesses can learn that pro-environment is not antigrowth.[11]

The thirtieth anniversary of Earth Day, celebrated on April 22, 2000, on the mall in Washington, D.C., was remarkable as much for what did not happen as for what did. American activists accustomed to being resented and resisted found themselves frustrated by the realization that they had won the war on rhetoric and on many other fronts as well. They were like the dog that has been chasing the car for years and finally catches it, and does not have a clue about what to do. Environmental awareness is now embraced by the more enlightened factions of big business. Not just Ford but Toyota, Honda, and other major firms sponsored parts of the huge party and handed out information boasting of their efforts to do something important for the environment. The environmental movement is learning that it can no longer treat these people as the enemy. The movement needs them to show others that you can be a good corporate citizen and still make money; even more, environmentalists need their help. It is great that so many car companies have figured out how to market themselves as green. It will be even greater when they commit to producing and marketing cars that use less fossil fuels more efficiently, and ultimately, use none at all.

Half a million persons gathered on the mall in Washington, D.C., for an event that was webcast to half a billion people globally. Chevy Chase gave the best-attended press conference of the day, in which he deadpanned that though he knew a little bit more than his family dog on the subject, at least he could do a bit more than bark about it. Esaih Morales, the star of *Chicago Hope,* passionately pleaded with young people to get involved in global action and to treat their Mother Earth right.

I felt a bit like the skunk at the garden party. Denis Hayes, the organizer, had urged me to tell the story of Donora as a warning. On the Sabbath of Passover, I spoke on behalf of those who had not survived pollution that had never even been charted. I cautioned that most of the world's children today are growing up in third-world cities where unrelenting pollution is just a way of life. In many parts of the world the conditions that took place in Donora and London can occur and can be prevented. Even in our own country, we are still paying the price of failing to control some of the oldest, dirtiest sources of energy and failing to protect our poorest citizens from industrial wastes. We now understand that local pollution ultimately creates weather all over the world by affecting the fragile envelope of greenhouse gases that makes life possible on this planet. We honor the dead, I told my listeners, to warn the living.

In early 2002, I rushed to Pittsburgh. My mother had had a heart attack. After all these years, it still was a shock to hear those words. She has always seemed nearly immortal to me. I know that none of us lives forever. I am a scientist. I am also religious. Both traditions say emphatically that when our time comes, it comes. Still, I was not ready for that call.

My mother made it this time. She may well be on her way to joining a crowd of Donora's long-term survivors, a group who got through the dirtiest days of Donora and have lived well into their nineties—including Devra Miller Breslow's aunt and uncle, as well as my Aunt Gertrude and Uncle Harry. Some folks have proved to be resistant to pollution's effects, and it would be interesting to learn why.

After she came through this heart attack, my mother said to me, "Of course I made it. When I was five years old, I walked 105 steps four times a day back and forth to school in the morning and afternoon five days a week. We didn't have those streetcars like city people. We had to be strong."

"So you knew how to count that high at that age?" I asked.

"Sure," she replied. "I kept counting 'cause I wanted to be sure the number did not change. You had to be tough to live in Donora. All those steps every day."

A few years ago, when she was climbing steps better than she does now, my mother took my daughter, Lea, and niece Kate to an amusement park in Central Pennsylvania. She commanded them to ride all the rolly coasters in the place. At the first two, my mother sat and waited on a nearby bench, as her granddaughters would slowly swoop up—clack-a-clack-a-clack—and suddenly drop down and around all those fast, sudden, but safe curves. After she and Kate had completed their third ride, my daughter suggested that the three of them just sit and watch the rest of the coasters together.

"No!" my mother insisted. "You two, you go on, because you can. I used to love to ride rolly coasters. Now I just get to watch."

To this day, my mother does not believe that living in Donora had anything to do with her lifelong troubles with heart disease or the cardiac ailments of her four siblings. After her most recent heart attack, my siblings were sure she would be fine. They were right. The belief in medical miracles and the sense of individual responsibility for health runs deep as biblical guilt allows.

Public health studies can tell you what the chances are that a number of people in any given large population will grow sick and die from certain illnesses. But these studies cannot say why a particular individual succumbs. Can anybody say what role growing up in Donora may have played in my mother's heart disease and that of her four siblings? Why my beloved Uncle Len suddenly dropped dead at age fifty? Why my brother Stan and I, both of us aging jocks, have developed severe allergies, or why my dad, who began working in the steel mill at age fifteen, contracted bone cancer at age fifty-three? All we really know is that the chances of developing heart disease and cancer—and the chance of dying from them—remained high in Donora for more than a decade after the killer smog. As to allergies and asthma, rates are soaring in children and adults all over the world, and probably in those of us who grew up in Donora.

In my mother's survival, in her refusal to give in, and her joy in watching rolly coasters today, I see an eloquent form of courage—a defiance of her circumstances. The poet and writer Sandra Steingraber, a cancer survivor at age nineteen, long before she dreamed of having her beloved daughter Faith, inspires many others with her remarkable conviction in the possibility of transforming this world if only we accept the importance of the environment in determining our health. The unsinkable Andrea Martin, with the strength of personality of one of those Chinese dolls that keeps bouncing back, has begun a national movement to promote research on the environment and the development of precautionary actions to clean up our homes, schools, gardens, and workplaces.

Five years ago, after the Breast Cancer Fund began to focus on environmental issues, I stood with Susan Shinagawa, a fellow board member and a longtime cancer survivor, on a deck overlooking the Pacific Malibu coast. We spoke of the need to reach out and ride the waves before us, to keep going despite the difficult times, which for Susan were nearly overwhelming. At that moment, a school of dolphins surfaced, leaping and catching the waves in the ocean less than a hundred yards offshore. We felt as though we could have ridden them away.

I remembered a modern Midrashic story I learned as a child. A rabbi is walking along a beach at low tide, picking up stranded starfish one at a time and tossing them into the water. A little boy comes up to her and says, "Why are you wasting your time? Tomorrow the beach will be covered again with starfish. What you're doing won't make any difference."

"Oh, no?" the rabbi replies. She picks up a starfish and flings it far into the sea. "Made a difference to that one."

ACKNOWLEDGMENTS

WILLIAM FRUCHT, SENIOR EDITOR of Basic Books, consistently wheedled and prodded me in what appears to have been the right direction, as did Richard Morgenstern, my venerated husband of some 28 years. Fortunately for both of us, my husband shares my passions for this work and provided candid, constructive criticisms along with love and support to bear most of them. Throughout this process, my adventurous children, Aaron and Lea, have reliably offered leveling, good humor, great food, and much patience.

A number of people made the writing of this book considerably more challenging than it would otherwise have been, and I am indebted to all of them: Jerold Abraham, Rita Arditti, David Bates, Gordon Binder, Victor Borja-Aburto, H. Leon and Hattie Bradlow, Barbara Brenner, Devra Miller Breslow, Lester Breslow, Phil Brown, Susan Cantor, Barry Castleman, B. J. Croall, Elizabeth Dance, Nancy Evans, Robert Fri, Bernard Goldstein, Nelson Gouveia, Kim Hooper, James Huff, Matti Jantunen, Ted Kerstetter, John M. Last, Lester Lave, Barbara Lazarus, Phil Lee, Mort Lippmann, Amy Loomis, Robert Maynard, John McLachlan, Luisa T. and Mario Molina, Herbert Needleman, Cheryl Osimo, Harry Paxton, Bill Pedersen, Lucian Pugliarisi, Joan Reiss, Tom Roper, F. Sherwood Rowland, Carlos Santos-Burgoa, Anne Sasco, Yasmin von Schirnding, Ilene Schwartz, Allen Silverstone, Joseph Sommers, Jack Spengler, Paul Stolpman, Eric Stork, Guodong Sun, Joel Tarr,

Myron Taube, Daniel Teitelbaum, George Thurston, John C. Topping, Jr., and Daniel Wartenberg.

Hillary Stainthorpe provided research and editing skills far beyond her tender years for many parts of the process. Barbara Fitzgerald, Courtney Patterson, Kim Provenza, Linda Roberts, Elizabeth Topping, John Topping III, Nicole Burdick, and Guli Fager assisted in final preparations of this book, including more late night forays than any of us would care to admit. Tiffany Miles, Miranda Loh, Dina Penny, Ruchi Bhandari, Seema Paul, Changhua Wu, Angela Ip, Michelle Gottlieb, Judy Pongsiri, Ora Sheinson, Julie Stampnitzky, Darius Sivin, and Barbara Ley, worked on earlier parts of the research and have since gone on to greater glory and much more profitable endeavors.

Sean Beggs, Marie Coleman, Margie Hinebaugh, Jeffrey Hunker, Mark Kamlet, Barbara Lazarus, Emily Marshall, Indira Nair, and other colleagues of Carnegie Mellon University's H. John Heinz III School of Public Policy and Management and the Engineering and Public Policy Program provided an array of support and problem-solving for which I am deeply grateful.

Michelle Bell creatively and diligently developed much of the original investigations of London's killer smog of 1952–1953 on which Chapter Two relies. Luis Cifuentes, George Thurston, Nelson Gouveia, Binghen Chen, Paulo Saldiva, Victor Borja, Nasir Khattak, and A. Karim Ahmed, collaborated with me across many time zones and computer systems to create the air pollution analyses that are described here. Deborah Axelrod, H. Leon Bradlow, Mitchell Gaynor, Nitin Telang, Michael Osborne, and Jack Fishman have worked with me in developing research and testing theories on hormones and cancer. Rikuo Doi and Hillary Stainthorpe have generated some important analyses of time trends in Japan and the United States on sex ratio at birth.

For providing intellectual and spiritual support for my research, many really good arguments, and invaluable challenges over the years, I am indebted to my mother Jean Langer Davis, my mother-in-law Mildred G. Morgenstern, my sister, Sara Davis Buss, my brothers, Martin and Stanford Davis, and the late David Abramson, Deborah Axelrod, Olav Axelson, Harvey Babich, Bernice Balter, Livia and David Bardin, Dan and Carol Berger, Eula Bingham, Sharon Bloom, Susan Blumenthal, John Buss,

Catherine Cameron, Joseph Cannon, Susan Canter, Aaron Cohen, Audrey Chapman, Luis Cifuentes, R. F. Culbertson III, Ann Davis, Leonard, Marion, and Molly Davis, Joseph DeCola, Fred Dobbs, Douglas Dockery, Akiko Domoto, Joycelynn Elders, Mike Fitzgerald, Tony Fletcher, Harold Freeman, J. William Futrell, Mitchell Gaynor, Gang Ke, David Gee, Laurie Goodman, Daniel Greenbaum, Wade Greene, Andy Haines, John and Olivann Hobbie, David Hoel, Gary Hook, Arnold and Rina Hirsh, Yiping Hu, Peter Infante, Karen Folger Jacobs, Mark Jacobs, Beverly Jackson Jones, Lenore Kohlmeier, Amy Kyle, Richard Laermer, Phil Landrigan, Jonathan Lash, Ronnie Levin, Changsheng Li, Bonnie Lifton, Lizbeth Lopez, Doree Lynn, Amy Lyons, Henry Maddoff, Ed Markey, Cesare Maltoni, Andrea Martin, Gerald McCarthy, Phil and Adelle Morgenstern, Beth Morgenstern, David McNamee, Myron Mehlman, Anthony Miller, Avis Miller, Karen Miller, Victor Miller, Dale Mintz, Ronald and Cathy Muller, J. Peterson Myers, Kirsten Niblaeus, Hope Nemiroff, Michael Nussbaum, Cheryl Osimo, Jayne Ottman, John Pan, Paul Petro, Penelope Pereira, Andrew Polsky, Frank Press, the late David Rall, Noel Raskin, Elihu Richter, Ruthann Rudel, David Saperstein, the late Marvin Schneiderman, Bonnie Nelson Schwartz, Joel Schwartz, Jill and Richard Sheinbaum, Janette Sherman, Diane Shrier, Ellen Silbergeld, Lisa Simpson, Tiger Steuber, Warren Stone, Daniel Swartz, David Suzuki, Elizabeth Sword, Rafael Tarnopolsky, Sam Their, Lorenzo Tomatis, Jeffrey Tuckfelt, Mark and Sondra Tuckfelt, Yasmin von Schirnding, Jay Schulkin, David Walls-Kaufman, Sophia Wakefield, Gloria Weissberg, Robert White, Shawnna Willey, Jeff Wohlberg, Steve Wolin, Tracey Woodruff, and my agent, Peter Cox, of International Literary Management.

I have learned much from my colleagues at a number of institutions, some of which have directly provided support, including:

American Association for the Advancement of Science, Breast Cancer Fund, Adas Israel Synagogue's Havera, Carnegie Mellon University's H. John Heinz III School of Public Policy and Management, Children's Health and Environment Coalition, Climate Institute, Coalition of Organizations on the Environment and Jewish Life, Collegium Ramazzini, Department of Energy, Conservative Women's League, Environment Ministry of Sao Paulo, U.S. Environmental Protection Agency, European Environment Agency, Hadassah, Harvard University, Intergovernmental Panel on Climate Change, Johns Hopkins University, School of Public Health, Jennifer Altman Fund, Keep Yellowstone Nuclear Free, Susan B. Komen Foundation of Northern New Jersey, LocalMotion, London School of Hygiene and Tropical Medicine, Macalester College, Na amat, the National Institutes of Environmental Health Sciences, National Cancer Institute, National Institutes of Health, the National Religious Partnership on the Environment, National Renewable Energy Laboratory, Oberlin College, Environmental Studies Program, Pan American Health Organization, Rockefeller Brothers Fund, Rockefeller Family Financial Services, Silent Spring Institute, The Stern College of Yeshiva University, United Nations Development Program, United Nations Environment Program, W. Alton Jones Foundation, Wallace Genetics, Wallace Global Foundation, Women's Community Cancer Project, World Bank, World Health Organization, World Resources Institute.

I doubt that it would have been possible to complete this work in the days before the search engine of www.google.com, or without the good graces of the Library of Congress, Photographic Services Department, the National Library of Medicine, Carnegie Mellon University's library system, Jackson Hole Public Library, and access to the archives of the American Asssociation for the Advancement of Science, http://jstor.org. Those who setup and maintain EPA's web sites deserve medals, as do scores of its dedicated, under-recognized employees.

Permissions have been granted by *The Nation* to reprint material from: "A Bad Air Day," by David Corn, from the March 24, 1997, issue; and from "The Secret History of Lead," by Jamie Lincoln Kitman, from the March 20, 2000, issue; and by the Universal Press Syndicate to reprint the *Doonesbury* cartoon from July 1983.

Any errors in this work are either due to the ghosts in the machines that now dominate our lives or to my own faults. To the Holy One above us all, who has allowed me to reach this season and maintain hope despite the world's terrifying struggles, I give whatever thanks can be rendered by mere words alone.

Devra Davis
Teton Village, Wyoming
July 31, 2002

NOTES

Chapter 1: Where I Come From

1. Lynn Page Snyder, "The Death-Dealing Smog over Donora, Pennsylvania: Industrial Air Pollution, Public Health, and Federal Policy, 1915–1963" (Ph.D. diss., University of Pennsylvania, 1994), 71.
2. Berton Roueche, *Eleven Blue Men* (Boston: Little, Brown, and Company, 1953), 196.
3. Ibid., 215.
4. Snyder, "Death-Dealing Smog," 39.
5. George Waldbott, *Health Effects of Environmental Pollutants* (St. Louis: C.V. Mosby, 1978), 10.
6. Ibid., 11.
7. Snyder, "Death-Dealing Smog," 33.
8. Chris Bryson, "A Secret History of America's Worst Air Pollution Disaster," *Earth Island Journal* 13 (1998): 4, available at http://www.earthisland.org/eijournal/fall98/fe_fall98donora.html.
9. H. H. Schrenk et al., *Air Pollution in Donora, PA: Epidemiology of the Unusual Smog Episode of October 1948* (Washington, D.C.: Federal Security Agency, 1949), 45.
10. P. Sadtler, "Fluorine Gases in Atmosphere as Industrial Waste Blamed for Death and Chronic Poisoning of Donora and Webster, Pa., Inhabitants," *Chemical and Engineering News* 26 (1948):3692.
11. Schrenk, *Air Pollution in Donora,* 45.
12. Clarence A. Mills, "The Donora Episode," *Science* 111 (1950): 67.
13. Ibid.
14. Ibid., 68.

Chapter 2: The Phantom Epidemic

1. Sir Arthur Mitchell; Dr. Buchan, "Influenza and Weather of London in 1891," *Journal of the Scottish Meteorological Society* 9 (1893): 141. *Provided by Robert Maynard.*

2. See http://www.classicreader.com/read.php/sid.1/bookid.221/sec.1.

3. David Bates, *Environmental Health Risks and Public Policy: Decision Making in Free Societies* (Seattle: University of Washington Press, 1994), 7.

4. Peter Brimblecombe, *The Big Smoke: A History of Air Pollution in London Since Medieval Times* (London: Methuen, 1987), 7.

5. Ibid., 9.

6. See http://www.britannia.com/bios/wmlaud/execution.html.

7. www.La.utexas.edu/research/poltheory/james/blaste/index.html. *Counterblaste to Tobacco* originally published in London, 1604, reprinted in *Works of King James*, (London: Putnam & Sons, 1905).

8. Ibid., 40, 47.

9. John Evelyn, *Fumifugium, or the Inconvenience of the Aer and the Smoak of London Dissipated* (London, 1661). Available at http://www.astext.com/history/fumiguf.html.

10. Ibid.

11. Ibid.

12. See http://www.pepys.info/fire.html.

13. See http://www.angliacampus.com/education/fire/london/history/greatfir.htm.

14. Ibid.

15. Brimblecombe, *The Big Smoke,* 52.

16. John Graunt, *Natural and Political Observations made upon the Bills of Mortality* (London, 1662). Available at http://www.ac.wwu.edu/˜stephan/Graunt/13.html.

17. Peter Bernstein, *Against the Gods: The Remarkable Story of Risk* (New York: John Wiley, 1996), 57.

18. Ibid.

19. Ibid.

20. Bates, *Environmental Health Risks,* 7.

21. Ibid.

22. Ibid., 10.

23. Michelle Bell and Devra Lee Davis, "Reassessment of Lethal London Fog of 1952; Novel Indicators of Acute and Chronic Consequences of Acute Exposure to Air Pollution," *Environmental Health Perspectives,* June 2001, 391.

24. See http://www.squall.co.uk/climate2.html.

25. Ibid.

26. See http://www.britannia.com/history/nar20hist6.html.

27. F. Pearce, "Back to the Days of Deadly Smogs," *Nature,* December 5, 1992, 25–28.

28. See http://www.squall.co.uk/climate2.html.

29. George L. Waldbott, *Health Effects of Environmental Pollutants* (St. Louis: C.V. Mosby, 1978).

30. UK Minister of Health, Committee of Departmental Officers and Expert Advisers, "Mortality and Morbidity During the London Fog of December 1952," Report 95.

31. Bell and Davis, "Reassessment."

32. UK Minister of Health, "Mortality and Morbidity."

33. World Health Organization, *Rapport Epidemiologique et Demographique* 6, no. 8 (1953): 209–226.

Chapter 3: How to Become a Statistic

1. California Air Resources Board, "California's Air Quality History: Recent Events," available at http://www.arb.ca.gov/html/brochure/history.htm.

2. Steve Nadis and James J. MacKenzie, *Car Trouble* (Boston: Beacon Press, 1993), 5.

3. Bradford Snell, "The Streetcar Conspiracy: How General Motors Deliberately Destroyed Public Transit," available at http://www.lovearth.net/gmdeliberatelydestroyed.htm.

4. Ibid.

5. Ibid.

6. California Air Resources Board, "California's Air Quality History."

7. See http://www.ccohs.ca/oshanswers/chemicals/ld50.html.

8. Curtis Moore, "DuPont's Duplicity: Profiting at the Planet's Expense," *Multinational Monitor* 11 (March 1990).

9. Jamie Lincoln Kitman, "The Secret History of Lead," *Nation,* March 20, 2000. Available at http://past.thenation.com/issue/000320/0320kitman.shtml.

10. Herbert L. Needleman, "Clamped in a Straightjacket: The Insertion of Lead into Gasoline," *Environmental Research* 74 (1997): 97.

11. Ibid., 97–98.

12. Kitman, "Secret History of Lead."

13. It is an interesting twist of history that the building in which much of this book was written—Hamburg Hall of the H. John Heinz III School of Public Policy and Management at Carnegie Mellon University—had also been the site of the Bureau of Mines in Pittsburgh, where lead was tested in the 1920s.

14. Ibid.; Peter Montague, "History of Precaution," part 1, *Rachel's Environment & Health Weekly,* March 27, 1997, available at http://www.monitor.net/rachel/r539.html.

15. Needleman, "Clamped in a Straightjacket," 102.

16. East Asia and Pacific Environmental Initiative, "Removing Lead from Gasoline in Indonesia," activity page, updated April 7, 2002, available at http://eapei.home.att.net/Activities/2000/2000EPM-01.htm.

17. Mary O. Amdur and Leslie Silberman, "Direct Field Determination of Lead in Air," *AMA Archives Industrial Hygiene* 10 (1954): 152–157.

18. Mary Amdur and Philip Drinker, "The Inhalation of Sulphuric Acid Mist by Human Subjects," *Archives of Industrial Hygiene and Occupational Medicine,* 1952.

19. A character whose history appears to parallel Mary Amdur's appears in Saul Bellow's 1982 novel, *The Dean's December.* Bellow modeled Professor Sam Beach on the renowned geochemist Clair Patterson. When Patterson turned from reckoning the age of the earth and the moon to tracking the skyrocketing growth of lead pollution and its insidious effects on the brain, he was dumped by both the U.S. Public Health Service and the American Petroleum Institute in the 1960s. Members of the board of trustees at California Institute of Technology, where he had carried out his award-winning research for two decades, asked the chair of his department to fire him. In 1995, the year that he received the Tyler World Prize for Environmental Achievement, Patterson died of an asthma attack. In the novel, Beach fails to persuade the dean to shift from deep and abstract science to more practical applications that show the blight of lead poisoning for urban life. Of this resistance toward the real-world implications of science, Bellow wrote: "The genius of these evils was their ability to create zones of incomprehen-

sion. It was because they were so fully apparent that you couldn't see them." Cited in Herbert L. Needleman, "The Removal of Lead from Gasoline: Historical and Personal Reflections," *Environmental Research* 84 (2000): 23n.

20. Alice Hamilton to Mary Amdur, July 9, 1953, Mary Amdur Papers, private collection, courtesy of David Amdur.

21. California Air Resources Board, "California Air Quality History: Key Events," April 21, 2000, available at http://www.arb.ca.gov/html/brochure/history.htm.

22. Arie Haagen-Smit, "Air Conservation," *Science* 128, no. 3329 (1958): 869–878.

23. Ibid., 877.

24. Ruth Ann Bobrov, "Use of Plants as Biological Indicators of Smog in the Air of Los Angeles County," *Science* 121, no. 3145 (1955): 510–511.

25. R. C. Wanta and Howard E. Heggestad, "Occurrence of High Ozone Concentrations in the Air Near Metropolitan Washington," *Science* 130, no. 3367 (1959): 103–104.

26. Edmund G. "Pat" Brown [governor of California], inaugural address, January 5, 1959, available at http://www.governor.ca.gov/govsite/govsgallery/h/documents/inaugural_32.html.

27. John R. Goldsmith and Lester Breslow, "Epidemiological Aspects of Air Pollution," *Journal of Air Pollution Control Association* 9 (1959): 129–132.

28. Alfred C. Hexter and John R. Goldsmith, "Carbon Monoxide: Association of Community Air Pollution with Mortality," *Science* 172, no. 3980 (1971): 265–267.

Chapter 4: How the Game Is Played

1. "Guardian: Origins of the EPA," *EPA Historical Publications,* Spring 1992. Available at http://www.epa.gov/history/publications/origins5.htm.

2. P. Shabekoff, *A Fierce Green Fire: The US Environmental Movement Past and Present* (New York: Hill and Wang, 1993).

3. William H. Rodgers, *Environmental Law Air and Water* (St. Paul, Minn.: West), 243.

4. U.S. Environmental Protection Agency, "William D. Ruckelshaus: Oral History Interview," EPA 202-K-92-0003, January 1993, also available at http://www.epa.gov/history/publications/ruck/index.htm.

5. George Wilson, "Auto Pollution Hearing to Open," *Washington Post,* January 21, 1975.

6. U.S. Environmental Protection Agency, "William D. Ruckelshaus."

7. A. P. Loeb, "Case Study: The Surgeon General and Ethyl Gasoline, 1922–26" (Chicago: Illinois Institute of Technology, 1996), 5.

8. "Endanger . . . is not a standard prone to factual proof alone. Danger is a risk, and so must be decided by the assessment of risk." *Ethyl Corp. v. EPA*, 541 F. 2d1 D.C. Circuit Court of Appeals, 1976, *certiarari denied* 426 U.S. 941, 96 S. Ct 2663, 49L. Ed. 2d 394 (1976).

9. See, http://www.sierraclub.org/sierra/199807.LOLASP

10. Warren Winkelstein, Edward W. Davis, Charles S. Maneri, and William E. Mosher, "The Relationship of Air Pollution and Economic Status to Total Mortality and Selected Respiratory System Mortality in Men," *Archives of Environmental Health* 14, no. 1 (1967): 162.

11. William Haenzel, David B. Loveland, and Martin G. Sirken, "Lung Cancer Mortality as Related to Residence and Smoking Histories," pt. 1, "White Males," *Journal of*

the National Cancer Institute 28 (1962): 947; and William Haenzel and Karl Tauber, "Lung Cancer Mortality as Related to Residence and Smoking Histories," part 2, "White Females," *Journal of the National Cancer Institute* 32 (1964): 803.

12. Lester B. Lave and Eugene P. Seskin, "Air Pollution and Human Health," *Science,* August 21, 1970, 723–733.

13. Ralph C. Stahman, "History of Clean Air Act and Motor Vehicle Control," paper presented at a meeting of Society of Automotive Engineers, Warrendale, Pa., 1989; Delbert S. Barth, "Federal Motor Vehicle Emissions Goals for CO, HC and NOx Based on Desired Air Quality Levels," *Journal Air Pollution Control Association*, 1970, 20: 519–523.

14. Senate Subcommittee on Air and Water Pollution, Committee on Public Works, *Air Pollution Hearings* (Washington, D.C.: U.S. Government Printing Office, 1966); cited in George Waldbott, *Health Effects of Environmental Pollutants,* 2nd ed. (St. Louis: Mosby, 1978), 36.

15. Lester B. Lave and Eugene P. Seskin, *Air Pollution and Human Health* (Baltimore: Johns Hopkins University Press, 1977).

16. David Mastio, "Automakers Thrive Under EPA Rules: Early Predictions of Economic Damage Never Materialized," *Detroit News,* May 8, 2000, available at http://detnews.com/specialreports/2000/epa/auto/auto.htm.

17. Ibid.

18. Douglas Williams, "GM Proves EPA Right, Aide Says," *Detroit Free Press,* June 19, 1973.

19. Richard Wilson and Jack Spengler et al., *Health Effects of Fossil Fuel Burning* (Cambridge, Mass.: Ballinger, 1980).

Part 2: The Best of Intentions

1. S. J. Gould, "The Passion of Antoine Lavoisier," in *Bully for Brontosaurus: Reflections in Natural History* (New York: W. W. Norton, 1991), chap. 24; http://www.mun.ca/sgs/science/may2082.html.

Chapter 5: Zones of Incomprehension

1. Toby E. Huff, *Max Weber and the Methodology of the Social Sciences* (New Brunswick, N.J.: Transaction, 1984); and Jay Ciaffa, *Max Weber and the Problems of Value-Free Social Science: A Critical Examination of the Werturteilsstreit* (London: Associated University Press, 1998).

2. Sandra Harding, *Whose Science, Whose Knowledge?* (Ithaca, N.Y.: Cornell University Press, 1991), 3.

3. National Academy of Sciences, National Research Council, *Recommendations for the Prevention of Lead Poisoning in Children* (Washington, D.C.: NAS, 1976). Available from NTIS, Springfield, VA; PB 257645.

4. Herbert L. Needleman, "Lead Levels and Children's Psychologic Performance," *New England Journal of Medicine,* July 19, 1979, 163.

5. Herbert L. Needleman, "Childhood Lead Poisoning: The Promise and Abandonment of Primary Prevention," *American Journal of Public Health* 88, no. 12 (1998): 1871–1877.

6. Stephen Burd, "Scientists See Big Business on the Offensive," *Chronicle of Higher Education,* December 14, 1994, available at http://chronicle.com/data/articles.dir/art-41.dir/issue-16.dir/16a02601.htm.

7. Herbert Needleman, *Pediatrics* 90 (1992).

8. S. Cummings and L. Goldman, "Even Advantaged Children Show Cognitive Defects from Low-Level Lead Toxicity," *Pediatrics* 90 (1992): 995–997.

9. Herbert L. Needleman et al., "Bone Lead Levels and Delinquent Behavior," *Journal of the American Medical Association,* February 7, 1996, 363–369. See also Terrie E. Moffitt, "Measuring Children's Antisocial Behaviors," *Journal of the American Medical Association*, February 7, 1996, 403–404; and B. Bower, "Excess Lead Linked to Boys' Delinquency," *Science News*, February 10, 1996, 86.

10. John C. Stauber and Sheldon Rampton, "Bypassing Barriers with 'Active' and 'Passive' Public Relations," *PR Watch* 2, no. 3 (1995), available at http://www.prwatch.org/prwissues/1995Q3/bypass.html.

11. J. L. Annest, J. L. Pirkle, D. Makuc, J. W. Neese, D. D. Bayse, and M. G. Kovar, "Chronological Trend in Blood Lead Levels between 1976 and 1980," *New England Journal of Medicine* 308 (1983): 1373–1377.

12. Joel Schwartz and Hugh Pitcher, "The Relationship between Gasoline Lead and Blood Lead in the United States," *Journal of Official Statistics* 5, no.4 (1989).

13. Albert L. Nichols, "Lead in Gasoline," in *Economic Analyses at EPA,* ed. Richard Morgenstern (Washington, D.C.: Resources for the Future, 1997), 74.

14. Kitman, "Secret History of Lead."

15. John R. Goldsmith and Alfred C. Hexter, "Respiratory Exposure to Lead: Epidemiological and Experimental Dose-Response Relationships," *Science*, new series, October 6, 1967, 132–134.

16. Robert A. Kehoe, in "Proceedings of the Fifteenth Congress of Internal Medicine," *Vienna* 3, no. 3 (1966). Cited in Goldsmith and Hexter, "Respiratory Exposure," 134.

17. Kitman, "Secret History of Lead"; Nichols, "Lead in Gasoline."

18. Nichols, "Lead in Gasoline."

19. Sometimes local studies can be subject to especially heavy-handed pressure. Ingham County Health Department in Lansing, Michigan, saw an ill-named right-to-know study of its groundwater, its food, and growing rates of asthma in African American children shelved until Public Employees for Environmental Responsibility (PEER), a watchdog group, posted the results. See http://www.peer.org/michigan/index.html; and http://www.ecocenter.org/200112/lansing.shtml.

20. B. Weiser, "Forging a 'Convenant of Silence,'" *Washington Post,* March 13, 1989.

21. See http://www.hsia.org.

22. Daniel Wartenberg, D. Reyner, and C. S. Scott, "Trichloroethylene and Cancer: Epidemiologic Evidence," *Environmental Health Perspectives* 108, suppl. 2 (2000): 161–176.

23. International Institute of Synthetic Rubber Producers, "Unlike World War II, Nations Won't Face Rubber Shortage," press release, October 21, 2001, available at http://www.iisrp.com/press-releases/2001-Press-Releases/no-rubber-shortage.html.

Robert J. Lancashire, *Chemical Production Figures for the USA*, available at http://www.uwimona.edu.jm.1104/lectures/top50old.html. While U.S. production of butadiene has not increased much in the past decade, imports continue to grow.

24. J. E. Huff, R. L. Melnick, H. A. Solleveld, J. K. Haseman, M. Powers, and R. A. Miller, "Multiple Organ Carcinogenicity of 1,3-Butadiene in B6C3F1 Mice after 60 Weeks of Inhalation Exposure," *Science* 227 (1985): 548–549; and P. E. Owen, J. R. Glaister, I. F. Gaunt, and D. H. Pullinger, "Inhalation Toxicity Studies with 1,3-Butadiene," part 3, "Two-Year Toxicity/Carcinogenicity Study in Rats," *American Industrial Hygiene Association Journal* 48 (1987): 407–413.

25. Devra Davis, "Occupational Exposure to 1,3-Butadiene," *Lancet,* February 7, 1987.

26. U.S. Department of Health and Human Services, Public Health Service, National Toxicology Program, "1,3-Butadiene (CAS No. 106-99-00," in *Ninth Report on Carcinogens,* revised January 2001, available at http://ehp.niehs.nih.gov/roc/ninth/known/1_3_butadiene.pdf.

27. Lorenzo Tomatis, "The IARC Monographs Program: Changing Attitudes towards Public Health," *International Journal of Occupational and Environmental Health,* April-June 2002, 144–152.

28. Ibid.

29. Austin Bradford Hill, "The Environment and Disease: Association or Causation?" in *Report to the Workers' Compensation Board on Lung Cancer in the Hardrock Mining Industry,* Report 12 of Canada Industrial Disease Standards Panel (Toronto, Ontario: IDSP, March 1994), appendix, available at http://www.canoshweb.org/odp/html/rpt12.htm#APP-A.

30. Ibid.

31. J. I. Wallace, F. S. Rimm, H. Lane, H. Levin, E. L. Reinherz, S. F. Schlossman, and J. Sonnabed, "T-cell Ratios in Homosexuals," *Lancet,* April 1, 1982, 908.

32. M. L. Taff, F. P. Siegal, S. A. Geller, "Outbreak of an Acquired Immunodeficiency Syndrome Associated with Opportunistic Infections and Kaposi's Sarcoma in Male Homosexuals: An Epidemic with Forensic Implications," *American Journal of Forensic Medicine and Pathology* 3, no. 3 (1982): 259–264.

33. Michael Grunwald, "Monsanto Held Liable for PCB Dumping," *Washington Post,* February 23, 2002.

34. See http://www.greenpeaceusa.org/toxics/canceralleytour/ibervilletext.htm.

35. D. W. Dockery, C. A. Pope III, J. D. Spengler, J. H. Ware, M. E. Fay, Ben Ferris, and Frank Speizer, "An Association Between Air Pollution and Mortality in Six U.S. Cities," *New England Journal of Medicine,* December 8, 1993, pp. 1753–1759.

36. Wilson and Spengler, "Particles in the Air."

37. David Corn, "A Bad Air Day," *Nation,* March 24, 1997.

38. Arden Pope et al., "Lung Cancer, Cardiopulmonary Mortality, and Long-Term Exposure to Fine Particulate Air Pollution," *Journal of the American Medical Association,* March 2002.

39. Winkelstein, et. al., "The Relationship of Air Pollution and Economic Status to Total Mortality and Selected Respiratory System Mortality in Men, pt. I: Suspended Particulates," *Archives of Environmental Health* 14(1967): 162.

40. Stocks, P., "Lung Cancer and Bronchitis in Relation to Cigarette Smoking and Fuel Consumption in Twenty Countries," *British Journal of Preventative and Social Medicine* 21(1967): 181; Stocks, P., and J. M. Campbell, "Lung Cancer Death Rates Among Nonsmokers and Pipe and Cigarette Smokers: An Evaluation in Relation to Air Pollution by Benzopyrene and Other Substances," *British Medical Journal* 2(1955): 923.

41. Hammond, E. C., and D. Horn, "Smoking and Death Rates: Report on Forty-four Months of Follow-up of 187,783 Men, Pt. II: Death Rates by Cause," *Journal of the American Medical Association* 166(1958): 1294.

Chapter 6: The New Sisterhood of Breast Cancer

1. S. D. Stellman and Q. S. Wang, "Cancer Mortality in Chinese Immigrants to New York City: Comparison with Chinese in Tianjin and with White Americans," *Cancer* 73 (1994): 1270–1275.

2. L. A. G. Ries et al., eds., *SEER Cancer Statistics Review, 1973–98: Tables and Graphs* (Bethesda, MD: SEER, 2001), available at http://seer.cancer.gov/Publications/CSR1973_1998.

3. G. E. Dinse and D. M. Umbach et al., "Unexplained Increases in Cancer Incidence in the United States from 1975 to 1994: Possible Sentinel Health Indicators?" *Annual Review of Public Health* 20 (1999): 173–209.

4. World Health Organization, *The World Health Report 1997: Conquering Suffering—Enriching Humanity* (Geneva: World Health Organization, 1997); D. M. Parkin et al., *Cancer Incidence in Five Continents,* vol. 7, International Agency for Research on Cancer Scientific Publication 143 (Lyon: IARC, 1997); and M. P. Coleman et al., eds., *Trends in Cancer Incidence and Mortality,* IARC Scientific Publication 121 (Lyon: IARC, 1993).

5. R. Peto et al., "UK and USA Breast Cancer Deaths Down 25% in Year 2000 at Ages 20–69 Years," *Lancet* 355 (2000): 1822–1830.

6. American Cancer Society, "Cancer Facts and Figures: 1999," information brochure published by American Cancer Society, available at http://www.cancer.org.

7. Susan Ferraro, "The Anguished Politics of Breast Cancer," *New York Times Magazine,* August 15, 1993, cover and article, as discussed in Matushka Archive, Web page available at http://mchip00.nyu.edu/lit-med/lit-med-db/webdocs/webart/matuschka2-art-.html.

8. John Wargo, *Our Children's Toxic Legacy: How Science and Law Fail to Protect Us from Pesticides* (New Haven, CT: Yale University Press, 1998).

9. Rachel Carson, *Silent Spring* (Boston: Houghton Mifflin, 1962).

10. Gino J. Marco, Robert M. Hollingworth, and William Durham, eds., *Silent Spring Revisited* (Washington, D.C.: American Chemical Society, 1986).

11. Sandra Steingraber, *Having Faith* (New York: Basic Books, 2001).

12. Roy Wolford, quoted in Charlene Laino, "Eat Less, Live Longer," *Sightings* Web site, January 22, 1999, available at http://www.rense.com/health/eatless.htm.

13. D. Smith, "Worldwide Trends in DDT Levels in Human Breast Milk," *International Journal of Epidemiology* 28 (1999): 179–188.

14. Suzanne M. Snedeker, "Pesticides and Breast Cancer Risk: A Review of DDT, DDE, and Dieldrin," *Environmental Health Perspectives* 109, supplement 1 (2001): 35–47.

15. J. C. Semenza et al., "Reproductive Toxins and Alligator Abnormalities in Lake Apopka, Florida," *Environmental Health Perspectives* 105 (1997): 1030–1032 (1997).

16. T. B. Hayes et al., "Hermaphroditic, Demasculinized Frogs after Exposure to the Herbicide Atrazine at Low Ecologically Relevant Doses," *Proceedings of the National Academy of Sciences*, April 16, 2002, 5476–5480. Abstract available at http://www.pnas.org/cgi/content/abstract/99/ 8/5476.

17. D. L. Davis et al., "Rethinking Breast Cancer Risk and the Environment: The Case for the Precautionary Principle," *Environmental Health Perspectives* 106 (1998):

523–529; and W. B. Grant, "An Ecologic Study of Dietary and Solar UV-B Links to Breast Cancer Mortality Rates," *Cancer* 94 (2002): 272–281.

18. L. A. Brinton et al., "Breast Cancer Risk Among Women Under Fifty-Five Years of Age by Joint Effects of Usage of Oral Contraceptives and Hormone Replacement Therapy," *Menopause* 5, no. 3 (1998): 145–151.

19. G. T. Beatson, "On the Treatment of Inoperable Cases of Carcinoma of the Mama: Suggestions for New Methods of Treatment with Illustrative Cases," *Lancet* 2 (1896): 102-107, 162–165.

20. Jennifer L. Kelsey and Leslie Bernstein, "Epidemiology and Prevention of Breast Cancer," *Annual Review of Public Health* 17 (1996): 47–67.

21. D. L. Davis, G. Friedler, D. Mattison, and R. Morris, "Male-Mediated Teratogenesis and Other Reproductive Effects: Biologic and Epidemiologic Findings and a Plea for Clinical Research," *Reproductive Toxicology* 6 (1992): 289–292.

22. Committee on the Biological Effects of Ionizing Radiation (BEIR), Board on Radiation Effects Research, *Health Effects of Exposure to Low Levels of Ionizing Radiation: BEIR V* (Washington, D.C.: National Academy Press, 1989), 421.

23. A. Ekbom et al., "Intrauterine Environment and Breast Cancer Risk in Women: A Population-Based Study," *Journal of the National Cancer Institute* 89 (1997): 71–76.

24. C. C. Hsieh et al., "Twin Membership and Breast Cancer Risk," *American Journal of Epidemiology* 136 (1992): 1321–1326.

25. D. C. Malins, E. H. Holmes, N. L. Polissar, and S. J. Gunselman, "The Etiology of Breast Cancer: Characteristic Alterations in Hydroxyl Radical-Induced DNA Base Lesions During Oncogenesis with Potential for Evaluating Incidence Risk," *Cancer* 71 (1993): 3036–3043.

26. A. M. Soto et al., "P-Nonyl-Phenol: An Estrogenic Xenobiotic Released from 'Modified' Polystyrene," *Environmental Health Perspectives* 92 (1991): 167–173.

27. H. L. Bradlow, R. J. Hershcopf, C. P. Martucci, and J. Fishman, "Estradiol 16-Hydroxylation in the Mouse Correlates with Mammary Tumor Incidence and Presence of Murine Mammary Tumor Virus: A Possible Model for Hormonal Etiology of Breast Cancer in Humans," *Proceedings of the National Academy of Sciences, USA* 82 (1985): 6295–6299.

28. H. L. Bradlow et al., "Effects of Pesticides on the Ratio of 16a/2-Hydroxyestrone: A Biologic Marker of Breast Cancer Risk," *Environmental Health Perspectives* 103, supplement 7 (1995): 147–150.

29. T. Colton et al., "Breast Cancer in Mothers Prescribed Diethylstylbestrol in Pregnancy: Further Follow-Up," *Journal of the American Medical Association* 269 (1993): 2096–2200.

30. International Joint Commission, *Seventh Biennial Report Under the Great Lakes Water Quality Agreement of 1978* (Washington, D.C., and Ottawa, Ontario: International Joint Commission, 1994), chapter 3.

31. American Public Health Association, *Recognizing and Addressing the Environmental and Occupational Health Problems Posed by Chlorinated Organic Chemicals,* Policy Statement 9304 (Washington, D.C.: American Public Health Association, 1993).

32. M. I. New, "Premature Thelarche and Estrogen Intoxication," in *Estrogens and the Environment,* ed. J. A. McLachlan (New York: Elsevier Science, 1985).

33. I. Colon, "Identification of Phthalate Esters in the Serum of Young Puerto Rican Girls with Premature Breast Development," *Environmental Health Perspectives* 108 (2000): 895–900.

34. B. C. Blount, M. J. Silva, and S. P. Caudill, "Levels of Seven Urinary Phthalate Metabolites in a Human Reference Population," *Environmental Health Perspectives* 108 (2000): 979–982.

35. S. Pagan, "Early Puberty Linked to Shampoos," *New Scientist,* April 6, 2002.

36. Barbara Brenner, "Sister Support: Women Create the Breast Cancer Movement."

37. Breast Cancer Action's Web site is www.bcaction.org.

38. The Breast Cancer Fund's Web site is www.breastcancerfund.org.

39. D. Fagin, "A $30 Million Federal Study of Breast Cancer and Pollution on Long Island Has Disappointed Activists and Scientists Alike," *Newsday,* July 28, 2002.

40. Silent Spring Institute, Summary of Newton Breast Cancer Study, available at http://www.silentspring.org/Projects/Newton/Newton.html.

41. Janette Sherman, *Life's Delicate Balance: A Guide to Causes and Prevention of Breast Cancer* (New York: Taylor and Francis, 2000).

42. National Breast Cancer Coalition's Web site is www.natlbcc.org.

43. Many of the organizations discussed in these paragraphs have Web sites, some of which are listed here. Massachusetts Breast Cancer Coalition www.mbcc.org; Komen Foundation www.komen.org; National Organization for Women (NOW) www.now.org; Silent Spring Institute www.silentspring.org; Long Island Breast Cancer Coalition www.1in9.org; National Federation of Women's Institutes www.womens-institute.co.uk; and Women's Environmental Network www.wen.org.uk; Prevention Is the Cure preventionisthecure.org; Midhudson Options Project midhudsonoptionsproject.org.

44. Sherman, *Life's Delicate Balance.*

45. P. Muti, H. L. Bradlow, A. Micheli, V. Krogh, J. L. Freudenheim, H. J. Schunemann, M. Stanulla, J. Yang, D. W. Sepkovic, M. Trevisan, and F. Berrino, "Estrogen Metabolism and Risk of Breast Cancer: A Prospective Study of the 2:16alpha-Hydroxyestrone Ratio in Premenopausal and Postmenopausal Women," *Epidemiology,* November 2000, 635–640.

46. L. J. Vetten, B. O. Machle, T. I. Lund Nilson, S. Tretlin, C. C. Hsich, D. Trichopoulos, and S. O. Stuver, "Birth Weight as a Predictor of Breast Cancer: A Case-Control Study in Norway," *British Journal of Cancer,* January 7, 2002, 89–91.

47. P. A. Newcomb, B. E. Storer, M. P. Longnecker et al., "Lactation and a Reduced Risk of Premenopausal Breast Cancer," *New England Journal of Medicine,* January 13, 1994, 81–87.

48. R. A. Breslow, R. Ballard-Barbash, K. Munoz, and B. I. Graubard, "Long-Term Recreational Physical Activity and Breast Cancer in the National Health and Nutrition Examination Survey I Epidemiologic Follow-Up Study," *Cancer Epidemiology Biomarkers Preview,* July 2001, 805–808.

49. T. Kishida, M. Beppu, K. Nashiki, T. Izumi, and K. Ebihara, "Effect of Dietary Soy Isoflavone Aglycones on the Urinary 16-alpha to 2-Hydroxyestrone Ratio in C3H/HeJ Mice," *Nutrition and Cancer* 38, no. 2 (2000): 209–214.

50. G. Ursin, M. Wilson, B. E. Henderson, L. N. Kolonel, K. Monroe, H. P. Lee, A. Seow, M. C. Yu, F. Z. Stanczyk, and E. Gentzschein, "Do Urinary Estrogen Metabolites Reflect the Differences in Breast Cancer Risk Between Singapore Chinese and United States African American and White Women?" *Cancer Research,* April 15, 2001, 3326-3329.

51. J. Kliukiene, T. Tynes, and A. Andersen, "Risk of Breast Cancer Among Norwegian Women with Visual Impairment," *British Journal of Cancer,* February 2, 2001, 397–399.

52. E. S. Schernhammer, F. Laden, F. E. Speizer, W. C. Willett, D. J. Hunter, I. Kawachi, and G. A. Colditz, "Rotating Night Shifts and Risk of Breast Cancer in Women Participating in the Nurses' Health Study," *Journal of the National Cancer Institute,* October 17, 2001, 1563–1568.

53. Nancy Evans, *State of the Evidence: What Is the Connection Between Chemicals and Breast Cancer?* (San Francisco: Breast Cancer Fund/Breast Cancer Action, 2002).

Chapter 7: Save the Males

1. S. Meryn and A. R. Jadad, "The Future of Men and Their Health," *British Medical Journal* 323 (2001): 1013–1014.

2. Henrik Moller, "Trends in Sex-Ratio, Testicular Cancer, and Male Reproductive Hazards: Are They Connected?" *Acta Pathologica, Microbiologica et Immunologica Scandinavica* 106 (1998): 232–239; J. Toppari et al., "Male Reproductive Health and Environmental Xenoestrogens," *Environmental Health Perspectives* 104, supplement 4 (1996): 474–803.

3. Scientists at the U.S. Centers for Disease Control have no explanation for continuing increases in many birth defects in the past two decades, although improved recording certainly accounts for some of the growth. Larry D. Edmonds, "Temporal Trends in the Prevalence of Congenital Malformations at Birth Based on the Birth Defects Monitoring Program, United States, 1979–1987," *Morbidity and Mortality Weekly Report, CDC Surveillance Summaries,* December 1990, 19–23. Sudden changes in genes cannot account for the unexplained increases. See also Ted Schettler, Gina Solomon, Paul Burns, and Maria Valenti, *Generations at Risk: How Environmental Toxins May Affect Reproductive Health in Massachusetts* (Cambridge: Greater Boston Physicians for Social Responsibility, 1996).

4. The Trust for America's Health provides information on children's health, birth defects, and other health indicators at the county level for the entire United States; see http:healthyamericans.org. Through its Web site, www.scorecard.org, Environmental Defense allows citizens to find out how their zip code stacks up compared to the rest of the country with respect to some basic facts on air and water and wastes.

5. D. L. Davis et al., "Male-mediated Teratogenesis and Other Reproductive Effects: Biologic and Epidemiological Findings and a Plea for Clinical Research," *Reproductive Toxicology* 6 (1992): 289–292.

6. http://www.marchofdimes.org/healthLibrary/351.htm.

7. J. C. Robinson, *Toil and Toxics* (Berkeley: University of California Press, 1991), xv.

8. Mark A. Wolf, "Results of Repeated Exposure of Laboratory Animals to Various Concentrations of 1,2-Dibromo-3-Chloropropane," report, Dow Chemical Biochemical Research Laboratory Report File T3.5-68-3,4,5,6, July 23, 1958.

9. T. R. Torkelson et al., "Toxicologic Investigations of 1,2-Dibromo-3-chloropropane," *Toxicology Applied Pharmacology* 3 (1961): 545–559.

10. Karen Brown and Lori Ann Thrupp, "The Human Guinea Pigs of Rio Frio," *The Progressive,* April 1991, 28–30, cited in "Pesticide Hazard in Costa Rica" *TED*

(Trade Environment Database) Case Studies 4, no. 1 (January 1995), available at http://www.american.edu/ted/COSTPEST.HTM.

11. Torkelson, "Toxicologic Investigations," 545–559.

12. Robinson, *Toil and Toxics,* xvi.

13. Lori Ann Thrupp, "Sterilization of Workers from Pesticide Exposure: The Causes and Consequences of DBCP-Induced Damage in Costa Rica and Beyond," *International Journal of Health Service* 21, no.4 (1991): 731–757.

14. H. Babich, D. Davis, and G. Stozky, "Dibromochloropropane (DBCP): A Review," *Science of the Total Environment* 17, no. 3 (1981): 207–221.

15. G. Potashnik, I. Yanai-Inbar, M. I. Sacks, and R. Israeli, "Effect of Dibromochloropropane on Human Testicular Function," *Israeli Journal of Medical Science* 15, no. 5 (1979): 438–442.

16. G. Potashnik, "A Four-Year Reassessment of Workers with Dibromochloropropane Induced Testicular Dysfunction," *British Journal of Obstetrics and Gynaecology* 90, no. 6 (1983): 587.

17. D. Whorton et al., "Infertility in Male Pesticide Workers," *Lancet* 2, no. 8051 (1997): 1259–1261.

18. J. Fuller, *The Poison That Fell From the Sky* (New York: Random House, 1977), 7.

19. R. Ramondetta and A. Repossi, *Seveso Twenty Years After: From Dioxin to the Oak Wood* (Fondezione Lambardia per l'Ambiente, 1998).

20. The exception for abortions in Seveso was later used by those who argued successfully for the legalization of abortion under Italian law in 1978—a law that was later ratified by a popular vote in 1981.

21. J. K. Mitchell, *The Long Road to Recovery: Community Responses to Industrial Disaster* (New York: United Nations Univeristy Press, 1996), available at http://www.unu.edu/unupress/unupbooks/uu21le/uu21le0a.htm.

22. Steingraber, *Having Faith.*

23. D. Butlin, "The Placenta: A Literature Review," University of Reading School of Animal and Microbial Sciences, available at http://simba.rdg.ac.uk/DAVE/Lit%20review.html#1.

24. M. L. Tenchini et al., "A Comparative Cytogenetic Study on Cases of Induced Abortion in TCDD-Exposed and Nonexposed Women," *Environmental Mutagens* 5, no. 1 (1983): 73–85.

25. Ibid.

26. P. A. Bertazzi, et al., "Cancer Incidence in a Population Accidentally Exposed to 2,3, 7, 8-tetrachloldibenzo-para-dioxin," *Epidemiology* 4, 5 (September 1993).

27. See http://www.simba;7385.rdg.ac.uk/Dave/Lit%20review.html#1; P. A. Bertazzi et al., "Dioxin Exposure and Cancer Risk: A Fifteen-Year Mortality Study After the 'Seveso Accident,'" *Epidemiology* 8, no. 6 (1997): 646–652.

28. G. Remotti et al., "The Morphology of Early Trophoblast After Dioxin Poisoning in the Seveso Area," *Placenta* 2, no.1 (January-March 1981): 53–62.

29. B. Mocarelli et al., "Paternal Concentrations of Dioxin and Sex Ratios of Offspring," *Lancet,* May 2000, 355; May 27, 2000, 1858–1863.

30. A wide range of industry supported criticisms of Our Stolen Future can be found at http://www.heartland.org/suites/environment/chemicals3.htm.

31. D. Fagin and M. Lavelle tell the story of how industries have used a variety of public relations techniques to keep four chemicals—alachlor, atrazine, formaldehyde, and perchloroethylene—on the market. They report that about half of all officials who have regulated chemicals for Democratic or Republic administrations—end up working for the industries they regulate. See Fagin and Lavelle, *Toxic Deception: How the Chemical Industry Manipulates Science, Bends the Law, and Endangers Your Health* (New York, 1999).

32. For more information on these issues, see http://www.ourstolenfuture.org.

33. Janet Raloff reported on this work and related studies in *Science News,* January 22, 1994, available at http://www.sciencenews.org/sn–edpik/LS–8.htm. Elizabeth Carlsen et al., "Evidence for Decreasing Quality of Semen During the Past Fifty Years," *British Medical Journal* 305 (1992): 609-613; Jacques Auger et al., "Decline in Semen Quality Among Fertile Men in Paris During the Past Twenty Years," *New England Journal of Medicine,* February 2, 1995, 281–285; Lawrence Wright, "Silent Sperm," *New Yorker,* January 15, 1996; Richard M. Sharpe and Niels E. Skakkebaek, "Are Oestrogens Involved in Falling Sperm Counts and Disorders of the Male Reproductive Tract?" *Lancet,* May 29, 1993, 1392–1395.

34. The battle over what are the right data to look at to determine changes in sperm over time, which countries or types of sources should be included, and how best to interpret the information continues in the usual scientific forums and with considerable public relation attention from the environmental community and the chemical industry. S. Swan et al., "Declining Sperm Count Revisited: An Analysis of 101 Studies Published 1934–1996," *Environmental Health Perspectives* 108 (2000): 961–966; B. D. Acacio et al., "Evaluation of a Large Cohort of Men Presenting for a Screening Semen Analysis," *Fertility and Sterility* 73 (2000): 595–597.

35. "Environmental Working Group Reports That Chemical Industry Self-regulation Blocked Independent Science," available at http://www.chemicalindustryarchives.org/dirtysecrets/testing/1.asp.

36. A. Alberini and K. Segerson, "Assessing Voluntary Programs to Improve Environmental Quality," *Environmental Resources Economics* 22 (2002): 157–184.

37. C. Cranor, *Regulating Toxic Substances.* New York: Oxford University Press, 1993.

38. For practical advice on how to cope with a broad range of environmental health threats to children, see: P. Landrigan, H. Needleman, and M. Landrigan, *Raising Kids Toxic Free* (Rodale Press, 2001); Children's Health and Environment Coalition www.checnet.org; Environmental Working Group www.ewg.org; Health Care Without Harm, a site devoted to encouraging health care facilities to reduce their use of plastics, mercury, and other harmful agents www.noharm.org; Physicians for Social Responsibility www.psr.org; Rachel's Hazardous Waste Weekly www.monitor.net/rachel; Tulane University Center for Environmental Hormone Research, Environmental Estrogens and Other Hormones Web site www.som.tulane.edu/ecme/eehome; Center for Bioenvironmental Research at Tulane and Xavier Universities www.cbr.tulane.edu.

39. M. P. Longnecker et al., "Association Between Maternal Serum Concentration of the DDT Metabolite DDE and Preterm and Small-for-Gestational-Age Babies at Birth," *Lancet* 358 (2001): 110–114.

40. M. Maisonet et al., "Relation Between Ambient Air Pollution and Low Birth Weight in the Northeastern United States," *Environmental Health Perspectives,* supplement 3 (2001): 351–356.

41. J. A. Staessen et al., "Renal Function, Cytogenetic Measurements, and Sexual Development in Adolescents in Relation to Environmental Pollutants: A Feasibility Study of Biomarkers," *Lancet* 357 (2001): 1660–1669.

42. Lennart Hardell et al., "Occupational Exposure to Polyvinyl Chloride as a Risk Factor for Testicular Cancer Evaluated in a Case-Control Study," *International Journal of Cancer* 73 (1997): 828–830.

43. David Forman and Henrik Moller, "Testicular Cancer," *Cancer Surveys* 19–20: 323–341.

44. Erik Bendvold, "Semen Quality in Norwegian Men over a Twenty-Year Period," *International Journal of Fertility* 34 (1989): 401–404; Erik Bostofte et al., "Has the Fertility of Danish Men Declined Through the Years in Terms of Semen Quality? A Comparison of Semen Qualities between 1952 and 1972," *International Journal of Fertility* 2 (1983): 91–95.

45. Shanna H. Swan et al., "Have Sperm Densities Declined? A Reanalysis of the Global Trend Data," *Environmental Health Perspectives* 105 (1997): 1228–1232.

46. L. Welsh, "Effects of Exposure to Ethylene Glycol Ethers on Shipyard Painters," *American Journal of Industrial Medicine* 14, no. 5 (1988): 509–563.

47. R. Hauser et al., "Environmental Organochlorines and Semen Quality: Results of a Pilot Study," *Environmental Health Perspectives* 110, no. 3 (2002): 229–233.

48. J. Garcia-Rodriguez et al., "Exposure to Pesticides and Cryptorchidism: Geographical Evidence of a Possible Association," *Environmental Health Perspectives* 104, no. 10 (1996): 1090–1095.

49. S. S. Hosie et al., "Is There a Correlation Between Organochlorine Compounds and Undescended Testes?" *European Journal of Pediatric Surgery* 10 (2000): 304–309.

50. V. F. Garry, "Pesticide Appliers, Biocides, and Birth Defects in Rural Minnesota," *Environmental Health Perspectives* 104, no. 4 (1996): 394–399.

51. H. O. Dickinson and L. Parker, "Why Is the Sex Ratio Falling in England and Wales?" *Journal of Epidemiology and Community Health* 50 (1996): 227-230.

52. M. Sakamoto, A. Nakano, and H. Akagi, "Declining Minimata Male Birth Ratio Associated with Increased Male Fetal Death Due to Heavy Methylmercury Pollution," *Environmental Research* 89, no. 2 (2001): 92–98.

53. H. Needleman et al., "Bone Lead Levels and Delinquent Behavior," *Journal of American Medical Association,* February 1996, 363–369; B. Bower, "Excess Lead Linked to Boys' Delinquency," *Science News,* February 10, 1996, 86.

54. E. K. Ong and S. Glantz, "Constructing 'Sound Science' and 'Good Epidemiology': Tobacco, Lawyers, and the Public Relations Firms," *American Journal of Public Health* 91, no. 11 (November 2001): 1749–1757.

55. Ong and Glantz, "Constructing 'Sound Science.'"

56. Ibid.

57. Ibid.

58. D. L. Davis et al., "Reduced Ratio of Male to Female Births in Several Industrial Countries," *Journal of the American Medical Association* 279, no. 13 (1998): 1018–1023.

59. "Testicular Cancer and Cryptorchidism in Relation to Prenatal Factors," *Cancer Causes and Control,* November 8, 1999, 904–912.

60. Moller, "Trends in Sex-Ratio."

61. Available at http://www.junkscience.com/news/davis/html.

62. National Academy of Sciences, National Research Council, *Hormonally Active Agents in the Environment* (Washington, D.C.: National Academy Press), 4.

63. Alex Kirby, "Young Danes' Sperm Count Dips," *BBC News,* May 4, 2000, available at http://news.bbc.co.uk/hi/english/sci/tech/newsid_736000/736230.stm; Martineau et al., "Cancer in Wildlife, a Case Study: Beluga from the St. Lawrence Estuary, Quebec, Canada," *Environmental Health Perspectives*, March 2002, 285–292.

64. D. Martineau et al., "Cancer in Wildlife." N. Nuttall, "Pollutants Blamed for Dual Sex Polar Bears," *Times* (London) June 1, 1998.

65. As reported in *The Scientist,* Daniel Martineau found that about 20 percent of all dead beluga whales died of cancer. Available at http://www.the-scientist.com/yr2000/ict/research–001002.html.

66. "Polar Bears and PCBs," *Academic Press Daily Insight,* June 24, 1998, available at http://academicpress.com/inscight/06241998/graphb.htm.

67. J. A. Hoy, R. Hoy, D. Seba, and R. H. Kerstetter, "Genital Abnormalities in White-Tailed Deer *(Odocoileus virginianus)* in West Central Montana: Pesticide Exposure as a Possible Cause," *Journal of Experimental Biology,* 2002.

Chapter 8: Earthquakes and Spouting Bowls

1. "September 7, 1999, Athens, Greece Earthquake," article on Web page of EQE International, available at http://www.eqe.com/revamp/greece1.htm.

2. Luisa Tan Molina and Mario J. Molina, *Air Quality in Mexico Megacity: An Integrated Assessment* (Cambridge: MIT Press, 2000).

3. Jacobo Finkelman, "Phasing Out Leaded Gasoline Will Not End Lead Poisoning in Developing Countries," *Environmental Health Perspectives* 104, no. 1 (January 1996), available at http://ehpnet1.niehs.nih.gov/docs/1996/104-1/editorial.html.

4. Molina and Molina, *Air Quality.*

5. Victor Borja et al., "Ozone, Suspended Particulates, and Daily Mortality in Mexico City," *American Journal of Epidemiology* 145 (1997): 258-268.

6. Molina and Molina, *Air Quality.*

7. D. L. Davis, P. Saldiva et. al., *Urban Air Pollution Risks to Children: A Global Environmental Health Indicator,* Environmental Health Notes (Washington, D.C.: World Resources Institute, 1999).

8. V. Borja, "Ozone."

9. L. Cifuentes, V. Borja-Aburto, N. Gouveia, G. Thurston, and D. L. Davis, "Hidden Health Benefits of Greenhouse Gas Mitigation," *Science,* August 17, 2001, 1254–1259.

10. A. Alebic-Juretic, A. Frkovic, and D. Simic, "The Effect of Air Pollution on the Outcome of Pregnancies," *International Journal of Gynaecology and Obstetrics* 75, no. 3 (December 2001): 315–316.

11. Molina and Rowland won the Nobel Prize with Paul Crutzen; see http://www.nobel.se/chemistry/Laureates/1995.

12. Kate Doyle, "Mexico's New Freedom of Information Law," *National Security Archive,* June 10, 2002, available at http://www.gwu.edu/~nsarchiv/NSAEBB/NSAEBB68/.

13. Molina and Molina, *Air Quality.*

14. "The City with the Grittiest Air," *Albion Monitor* (Web newspaper), July 20, 1999, available at http://www.monitor.net/monitor/9907a/lanzhou.html.

15. M. S. Friedman, "Impact of Changes in Transportation and Commuting Behaviors During the 1996 Summer Olympic Games in Atlanta on Air Quality and Childhood Asthma," *Journal of the American Medical Association* 285 (2001): 897–905.

16. The UN Development Program report on wastewater in China, available at http://www.unchina.org/undp/press/html/resource.html.

17. The Greenpeace China Web site is available at http://www.greenpeace-china.org.hk/chi. See also, http://www.chinaenvironment.net.

18. On a weekly basis, *China Daily News* publishes relative rankings of its major cities for air pollution. Available at chinadaily.com.

19. Guodong Sun and H. Keith Florig, "Revealed Willingness-to-Pay for Reduction of Air Pollution Risks in Urban China Collaborators," report abstract available athttp://hdgc.epp.cmu.edu/projects/abstracts/china-wtp.html.

20. Robert Temple, with introduction by Joseph Needham, *The Genius of China* (New York and London: Simon and Schuster, 1986).

Chapter 9: A Grand Experiment

1. D. J. D. Price, *Big Science, Little Science* (New York: Columbia University Press, 1963).

2. R. Gelbspan, *The Heat Is On: The Climate Crisis, the Coverup, the Prescription* (New York: Perseus Books, 1998).

3. R. Benedick, *Ozone Diplomacy* (Cambridge: Harvard University Press, 1991).

4. Alliance for Responsible Atmospheric Policy Web site is available at http://www.arap.org/.

5. Paul Brodeur, "Annals of Chemistry," *New Yorker,* June 9, 1986, 83.

6. Sharon Roan, *Ozone Crisis: The Fifteen-Year Evolution of a Sudden Global Emergency* (New York: John Wiley, 1989).

7. James Hammett, "Stratospheric Ozone Depletion," in *Economic Analyses at EPA, Resources for the Future,* ed. R. D. Morgenstern (Washington, D.C., Resources for the Future, 1997).

8. Ibid.

9. Roan, *Ozone Crisis,* 146.

10. Brodeur, "Annals of Chemistry," 83.

11. Robert H. Mason, "The Ozone: Chlorofluorocarbon Controversy: The Evolution of Theoretical Science's Influence on United States Policy," paper submitted to Humanities Sufficiency Program, Worcester Polytechnic Institute, Worcester, Massachusetts, available at http://people.atg.com/~bob/papers/WPI/sufficiency.txt; Alan S. Miller and Irving M. Mintzer, *The Sky Is the Limit* (Washington, D.C.: World Resources Institute, 1986), 24.

12. For the use of carbon dioxide to force oil out of wells, see Energy Resources Center of the Earth Engineering Center of Columbia University Web site, available at http://www.seas.columbia.edu/earth/Energy.html.

13. http://climatechange.unep.net.

14. National Park Service, Outer Banks Group, "The Cape Hatteras Light Station Relocation Project Named Outstanding Civil Engineering Achievement, May 8, 2000," press release, available at http://www.nps.gov/caha/achievement.htm.

15. Lee Lane, "Who Says 'A' Must Say 'B,'" Americans for Equitable Climate Solutions Web site, April 29, 2002, available at http://www.aecs-inc.org.

16. Luis Cifuentes et al. "Assessing the Health Benefits of Urban Air Pollution Reductions Associated with Climate Change Mitigation (2000–2020): Santiago, São Paulo, Mexico City, and New York," *Lancet* 350 (1997): 1341–1349.

17. Junk Science Web site, http://www.junkscience.com/news2/jones3.htm.

18. Bert Metz, Ogunlade Davidson, Rob Swart, and Jiahua Pan, eds., *Climate Change 2001: Mitigation Contribution of Working Group III to the Third Assessment Report of the Intergovernmental Panel on Climate Change (IPCC)* (Cambridge: Cambridge University Press, 2001).

19. See http://www.nrel.gov/icap/pdfs/chile_report.pdf. Five hundred local governments, comprising 8 percent of global greenhouse gas emissions are devising specific strategies to lower emissions. See, http://www.iclei.org.

20. National Academy of Sciences, Transportation Research Board, "Effectiveness and Impact of Corporate Average Fuel Economy (CAFE) Standards" (Washington, D.C.: National Academy of Sciences Press, 2002), available at http://www.nap.edu/books/0309076013/html/.

21. Union of Concerned Scientists, "Common Sense on Climate Change: Practical Solutions to Global Warming," available at http://www.ucsusa.org/environment/solutions.html.

22. See http://www.coejl.org.

23. Green Mountain Energy Company's Web site is http://www.greenmountain.com.

24. Sam Wyly, speech to Montreux's Clean Energy Forum, Aspen, Colorado, September 1999, available at www.climate.org.

25. U.S. Environmental Protection Agency, "Benefits and Costs of the Clean Air Act," Final Report to Congress on Benefits and Costs of the Clean Air Act, 1990 to 2010, EPA document no. 410-R–99–001 and Final Report to Congress on Benefits and Costs of the Clean Air Act, 1970 to 1990, EPA document no. 410-R–97–002, available at http://www.epa.gov/oar/sect812.

26. B. Lomberg, *The Skeptical Environmentalist* (Cambridge: Cambridge University Press, 2001); see also http://www.gristmagazine.com/grist/books/links121201.asp; and http://www.lomborg.com/.

27. J. F. Rischard, *High Noon* (New York: Basic Books, 2002).

28. A. J. McMichael, *Planetary Overload* (Cambridge: Cambridge University Press, 1993).

29. Maurice Strong, *Where on Earth Are We Going?* (Toronto: Alfred A. Knopf, 2000).

30. "A Survey of Global Environment," *Economist,* July 6, 2002, 15.

31. Ray Kopp, Richard D. Morgenstern, William Pizer, and Michael Toman, *A Proposal for Credible Early Action in U.S. Climate Policy* (Washington, D.C.: Resources for the Future/Weathervane, 2001).

32. Richard D. Morgenstern, "Reducing Carbon Emissions and Limiting Costs," in John A. Riggs, ed., *U.S. Policy on Climate Change: What Next?* (Washington, D.C.: Aspen Institute, 2002), 165–176.

Chapter 10: Defiant Figures

1. Barbara Distel and Ruth Jakusch, *Concentration Camp Dachau 1933–1945*, translated by Jenniferi Vernon (Brussels: Comite International de Dachau, 1978), 213.

2. *Den Toten Zur Ehr, Den Lebenden Zur Mahnung* (For the dead an honor, for the living, a warning), inscription on The Defiant Inmate Statue, Dachau Concentration Camp Memorial.

3. M. Grunwald, "Senators Assail EPA on Ala. Cleanup," *Washington Post,* April 20, 2002, available at http://www.washingtonpost.com/ac2/wp-dyn?pagename=article&node=&contentId=A18136-2002Apr19.

4. E. D. Richter, C. Soskolne, J. LaDou, and T. Berman, "Whistleblowers in Environmental Science, Prevention of Suppression Bias and the Need for a Code of Action," *International Journal of Occupational and Environmental Health* 7 (2001): 68–71.

5. In 2000, the U.S. surgeon general estimated the costs of heart disease at more than $300 billion annually; see http://www.seniors.gov/articles/0201/heart-disease.html.

6. The growth in greener and cleaner household products is phenomenal. See the following websites: Seventh Generation (cleaners and paper products): http://www.seventhgen.com; Earth Friendly Products (cleaners): http://www.ecos.com; Oxi Clean and Orange Glo (as seen on TV): http://www.oxyclean-orange-glo-oxiclean.com.

7. For the children's health and environment coalition electronic portal to more than eighty other groups, see http://www.checnet.org, http://yosemite.epa.gov/ochp/ochpweb.nsf/homepage, http://www.cehn.org. Also see P. Landrigan, H. L. Needleman, and M. Landrigan, *Raising Healthy Children in a Toxic World* (New York: Rodale Press, 2002).

8. See http://www.context.org/ICLIB/IC28/Robert.htm.

9. M. Gaynor and J. Hickey, *Dr. Gaynor's Cancer Prevention Program* (New York: Kensington Publishing, 1999).

10. R. O'Donnell, D. Axelrod, and T. Semler, *Bosom Buddies: Lessons and Laughter on Breast Health* (New York: Time Warner, 1999).

11. See http://www.andrewweil.com/.

12. D. Shephardson, "Rouge Smokestacks Will Fall," *Detroit News*, March 21, 2002, available at http://detnews.com/2002/wayne/0203/25/d03–445709.htm.

INDEX